OCS Study
MMS 2005-012

Coastal Marine Institute

Potential Spatial and Temporal Vulnerability of Pelagic Fish Assemblages in the Gulf of Mexico to Surface Oil Spills Associated with Deepwater Petroleum Development

Authors

Mark C. Benfield
Richard F. Shaw

February 2005

Prepared under MMS Contract
14-35-00001-30660-19962
by
Department of Oceanography and Coastal Sciences
Coastal Fisheries Institute
Louisiana State University
Baton Rouge, Louisiana 70803

U.S. Department of the Interior
Minerals Management Service
Gulf of Mexico OCS Region

Cooperative Agreement
Coastal Marine Institute
Louisiana State University

DISCLAIMER

REPORT AVAILABILITY

Extra copies of the report may be obtained from the Public Information Office (Mail Stop 5043) at the following address:

U.S. Department of the Interior
Minerals Management Service
Gulf of Mexico OCS Region
Public Information Office (MS 5043)
1201 Elmwood Park Boulevard
New Orleans, Louisiana 70123-2394

Telephone: (504) 736-2519 or
1-800-200-GULF

CITATION

Suggested citation:

Benfield, M.C. and R.F. Shaw. 2005. Potential spatial and temporal vulnerability of pelagic fish assemblages in the Gulf of Mexico to surface oil spills associated with deepwater petroleum development. U.S. Dept. of the Interior, Minerals Management Service, Gulf of Mexico OCS Region, New Orleans, LA. OCS Study MMS 2005-012. xv+158 pp.

ABOUT THE COVER

Yellowfin tuna, such as this specimen, are abundant around deepwater platforms. The fish in the upper image was collected near the Brutus platform. Photo credit: Joe Malbrough Jr., Louisiana Universities Marine Consortium. The present study used distributional data from the National Marine Fisheries Longline Database (lower left panel) and other sources to predict the monthly distributions of yellowfin tuna (lower right panel) and other species in the deepwater region of the northern Gulf of Mexico.

ACKNOWLEDGMENTS

The authors are extremely grateful to Elizabeth Burroughs-Loden for her diligent pursuit of the literature and data relevant to this project. We appreciate the assistance of Mr. Mark McDuff, National Marine Fisheries Service (NMFS) for providing access to the SEAMAP dataset and Dr. Jean Cramer, NMFS for providing access to the longline database for the Gulf of Mexico. Dr. Richard Pitman, NMFS generously provided access to his flyingfish data. Ms. Kathy Lang, NMFS provided data on distributions of larval yellowfin tuna. We thank the NMFS Library Staff in Miami for their assistance with our literature searches.

TABLE OF CONTENTS

TABLE OF CONTENTS (continued)

LIST OF FIGURES

LIST OF TABLES

1   Executive Summary

Advances in deepwater drilling and production technologies have resulted in the seaward expansion of gas and petroleum platforms beyond the continental shelf in the northern Gulf of Mexico. With increased industrial activities in deeper water comes an increased risk of spills in the deepwater pelagic zone. This region provides habitat for a variety of fish species of considerable ecological, recreational and commercial importance. Compared with fish species inhabiting the inshore waters of the northern Gulf, relatively little is known about the distributions and ecology of these offshore taxa.

Target taxa were pelagic fishes including selected members of the *Sargassum* community. Pelagic species were considered of primary importance because: (1) most produce large numbers of small eggs with limited yolk reserves that hatch into larvae dependent on plankton in the near-surface waters for nutrition; (2) most fisheries target pelagic fish taxa; (3) oil is buoyant and will accumulate in the neustonic zone; (4) based on slicks formed by natural petroleum seeps, even oil released from near the bottom will likely rise to the surface; and (5) it is unlikely that there is sufficient information on the distributions of demersal or benthic fishes to make even a well-reasoned inference about their spatial and temporal distributions. Target species selected for review were bluefin tuna (*Thunnus thynnus*), yellowfin tuna (*T. albacares*), blackfin tuna (*T. atlanticus*), blue marlin (*Makaira nigricans*), white marlin (*Tetrapterus albidus*), wahoo (*Acanthocybium solanderi*), dolphin (*Coryphaena hippurus*), blue runner (*Caranx crysos*), spotfin flyingfish (*Cypselurus furcatus*), Atlantic flyingfish (*C. melanurus*), ocean sunfish (*Mola mola*), and selected members of the pelagic *Sargassum* community: sargassumfish (*Histrio histrio*), planehead filefish (*Monocanthus hispidus*), and tripletail (*Lobotes surinamensis*).

This study was undertaken to review the available information on the distributions of these pelagic fish species in order to predict the spatial and temporal distributions of larval, juvenile and adult life history stages within the surface waters over an area likely to experience increasing gas and petroleum extraction. The study area was defined as the waters over the 200-2000 m isobaths and in some areas deeper than 2000 m extending from 28 °N south to 26 °N latitude and extending from 96.4 °W to 84.3 °W. This generally rectangular region was divided into three zones (western, central, and eastern). An additional region defined by a triangle with its apex at 87 °W, 30 °N and base extending from 90.7 °W, 28 °N to 84.3 °W, 28 °N was designated the northern zone.

Our review of the literature consulted peer-reviewed and non-peer-reviewed literature as well as Internet resources. This report drew heavily from the National Marine Fisheries Service long-line database and the Southeast Area Monitoring and Assessment Program (SEAMAP) ichthyoplankton surveys. These datasets provided distributional data for adults and larvae, respectively, however, obtaining data on the distributions of juveniles proved to be highly problematic because they avoid planktonic sampling gear and are not captured in commercial fisheries. Consequently, this report focuses on larvae and adults.

For each species, we have summarized the available distributional data on a monthly basis and have attempted to predict the distributions of larvae, adults, and juveniles (when possible) within the study region. Companion software in the form of Microsoft Excel spreadsheets allow the user to query the data to obtain probable distributions within specific locations defined by their longitude and latitude. It is clear that for many taxa, substantial gaps exist in our understanding of their spatial and temporal distributions and we hope that this study will provide a starting point for other studies designed to extend our knowledge of these poorly understood taxa.

2 Introduction

Recent advances in deepwater drilling and deepwater platform technologies combined with new petroleum and natural gas discoveries on the outer continental shelf (OCS) and slope have accelerated platform deployments in previously unexploited waters of the northern Gulf of Mexico. In spite of preventative measures established by the petroleum industry and governmental regulatory agencies to prevent the accidental discharge of petroleum products, oil spills remain a statistical certainty (e.g., Price and Marshall, 1996).

As petroleum exploration expands into, and beyond the waters of the OCS, the potential exists, via accidental spills, to adversely impact pelagic recreational and commercial fisheries. Surface petroleum spills in pelagic waters of the OCS will primarily impact those species of fishes and crustaceans that inhabit the epipelagic zone of the open ocean. Members of this group include several species that command a high monetary and socio-economic value (e.g., tunas, wahoos, and billfishes), as well as ecologically important or indicator species (e.g., flying fishes, ocean sunfishes). Spills in the surface waters are also likely to impact floating *Sargassum* communities, which contain a diverse and often unique faunal assemblage of fishes and invertebrates and which also serve as important nursery habitats for many fishes belonging to the families Coryphaenidae, Carangidae, Pomacentridae, and Lobotidae.

Relatively little is known about the susceptibility of pelagic fishes from the Gulf of Mexico OCS to petrochemical spills. The magnitude of any impact will depend upon the spatial and temporal scale of the incident as well as the chemical properties of the spilled material. The spatial scale (location, depth and extent) of the spill combined with the temporal scale (timing and duration) will combine to determine the species and life history stages that are likely to be present in the impacted area. Unfortunately, information on the spatial and temporal distributions of pelagic fish stocks in the OCS is not readily available and is generally scattered throughout the peer-reviewed and non-peer-reviewed technical literature and databases. This study was undertaken to synthesize what is known about the spatio-temporal distribution patterns of selected pelagic fish species. We have attempted to provide an estimate of what life history stages of these target species are likely to be present within the OCS waters on a seasonal basis.

2.1 Study Area

At the inception of this study, most of the active leases within the deepwater zone of the OCS lay within the 200-2000 m isobaths. Both the Mississippi and De Soto Canyons are encompassed by this region, however, we felt that a strict selection of the area overlying the 200-2000 m depth range would be too restrictive a criterion because some active leases lay outside of that zone in deeper water, and data from the locations of commercial fishing vessels (Maul et al. 1984), suggested that fish species such as bluefin tuna range through this zone and further to the south into the central Gulf. Accordingly, we delineated a study region of the north central Gulf of Mexico that includes waters above the 200-2000 m isobaths south to 26 °N latitude. The study area was divided into four zones: western zone (96.4 °W–92.0 °W, 26.0 °N–28.0 °N), central zone (92.0 °W–88.0 °W, 26 °N–28 °N), eastern zone (88 °W–84.3 °W, 26.0 °N–28.0 °N), and a triangular northern zone with a base from 90.7 °W–84.3 °W at 28 °N, and an apex at 87 °W, 30 °N (Fig. 1). The western, central and eastern zones correspond broadly to the MMS western, central and eastern planning areas.

While we are aware that the substantial network of planned and existing pipelines and subsea facilities creates a potential for deep-sea petroleum spills, we chose to focus on taxa that inhabit the upper 50 m of the water column (e.g., depth of the mixed layer) for the following reasons:

1. Most of our target species are highly fecund and produce large quantities of small eggs with limited yolk reserves. The small larvae that hatch from such eggs are dependent on the plankton for food and must forage in the near-surface waters;

2. Few commercially important demersal or benthic fisheries resources are currently being exploited in the deepest zones of the waters of the OCS;

3. Oil spilled in surface waters is buoyant and will likely accumulate in the neustonic zone;

4. Oil released near the bottom under high pressure and low temperature may remain at depth or rise to the surface depending upon the pressure, temperature, and viscosity of the oil. Naturally occurring surface slicks from deep-sea seeps are a common feature of the Gulf of Mexico suggesting that oil spilled near the bottom will likely rise to the surface; and

5. It is unlikely that sufficient information exists on the spatial and temporal abundances of deep-sea fishes to make even a well-reasoned inference about the impacts of subsea spills.

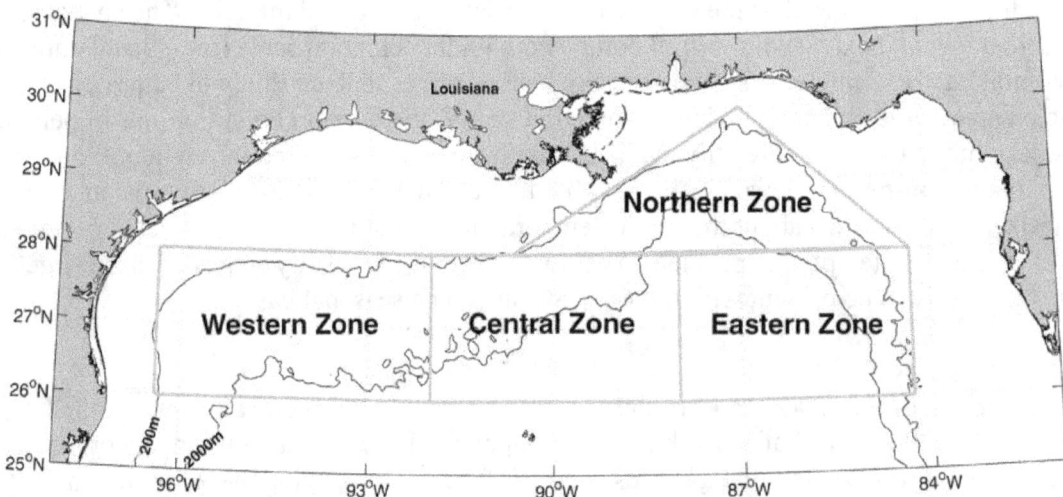

Figure 1. Locations of the western, central, eastern and northern study zones within which we attempted to define the distributions of target species.

2.2 Target Species

Bluefin tuna (*Thunnus thynnus*) are a large, long-lived species that represents the single most valuable fisheries resource on a value per pound basis. Demand for the raw flesh of this fish in the Japanese sashimi market has driven the price of individual fish to over $US 350 per pound (Safina 1993). Fishing pressure for bluefin tuna is intense and the western Atlantic breeding population appeared to drop by over 90% between 1975 and 1990 (Safina 1993). [At the time of writing this report, the theory that there are two distinct and separate stocks of bluefin tuna in the Atlantic was a subject of debate.] A long-lining fleet operates in the OCS deepwater region from approximately January to April (Maul et al. 1984). The Gulf of Mexico appears to be the only spawning ground for the western Atlantic stock of bluefin tuna (Clay 1991).

Yellowfin tuna (*T. albacares*) are the second most abundant true tuna in the Gulf of Mexico and represent a valuable commercial and recreational fishery resource that is targeted for human consumption. This is the secondary target species for the domestic and foreign bluefin tuna long-liner fleet and is a popular gamefish for recreational anglers.

Blackfin tuna (*T. atlanticus*) are the most abundant true tuna in the Gulf of Mexico where they are targeted as a popular recreational and commercial resource for human consumption. Blackfin tunas are frequently collected near petroleum platforms on the landward edge of the OCS.

Blue marlin (*Makaira nigricans*) and **white marlin** (*Tetrapturus albidus*) are highly sought after gamefish in the northern Gulf of Mexico.

Wahoo (*Acanthocybium solanderi*) are a popular gamefish in the OCS.

Dolphin (*Coryphaena hippurus*) also known as mahi mahi is a popular gamefish that is frequently targeted by recreational and commercial anglers. This fish is often found in close association with *Sargassum* rafts, weedlines and flotsam. The same processes that result in the accumulation of flotsam and weedlines may also cause aggregations of oil and tar.

Flyingfishes are potentially important indicator species because of their pelagic distribution, close association with the ocean surface, and their role as prey for many other larger predatory species. Two common species in the northern Gulf are *Cypselurus melanurus* and *C. furcatus* (J. Caruso, University of New Orleans, Pers Comm.).

Blue runner (*Caranx crysos*) are small- to medium-sized schooling carangids that are common prey items for larger predators. These fish are frequently associated with offshore petroleum platforms. Although humans seldom eat them, they are popular light tackle gamefish and are commonly captured for use as live bait.

***Sargassum* community fauna:** Two species of pelagic brown algae (*Sargassum fluitans* and *S. natans*) commonly called Gulf weed, form dense floating rafts in the pelagic waters of the Gulf. These rafts reproduce vegetatively and can occupy large areas covering hundreds to thousands of square meters. Within the *Sargassum* are a variety of fishes and invertebrates adapted to life in the weed through cryptic coloration, morphology and behavior. Common residents include the sargassumfish (*Histrio histrio*), sargassum pipefish (*Syngnathus pelagicus*), planehead filefish (*Monocanthus hispidus*), sargassum triggerfish (*Xanthichthys ringens*), and a variety of invertebrates including shrimp, nudibranchs, hydrozoans, and bryozoans. Rafts of *Sargassum* provide floating nurseries for juvenile carangids, sergeant majors (*Abudefduf saxatilis*) and tripletails (*Lobotes surinamensis*). *Sargassum* communities have the potential to be heavily impacted by spilled oil because patches of *Sargassum* often accumulate in the same areas where physical oceanographic processes are likely to concentrate oil.

3   Sources of Data

Surprisingly little information is available on the spatio-temporal patterns of abundance of commercially and ecologically important pelagic fishes in the waters beyond the shelf-slope break in the northern Gulf of Mexico. Records of the distributions of target taxa are scattered throughout the peer-reviewed and non-peer-reviewed 'gray' literature. In addition to a comprehensive literature search of the available peer-reviewed literature, gray literature (Table 1), and Internet websites, this report drew heavily on two datasets: the National Marine Fisheries Service (NMFS) long-line database and the Southeast Area Monitoring and Assessment Program (SEAMAP) ichthyoplankton surveys.

Table 1. Sources Consulted during the Literature Search for This Study.

Indices and Databases
Biological Abstracts
Cambridge Scientific Abstracts
Aquatic Sciences and Fisheries Abstracts
Conference Papers Index
Ecology Abstracts
Environmental Science and Pollution Management
Digests of Environmental Impact Statements
Oceanic Abstracts
Pollution Abstracts
Current Contents
Dissertation Abstracts
General Science Abstracts
National Technical Information Service Database
Science Citation Index/Web of Science
Uncover
Zoological Record

Library Collections
Louisiana State University
Rosenstiel School Library, University of Miami
National Marine Fisheries Service, Southeast Fisheries Center Library
NOAA Miami Regional Library
Louisiana Universities Marine Consortium Library

The NMFS long-line database contains the reported locations of the ends of surface long-line sets in the Gulf of Mexico. Originally comprised of Japanese vessels targeting adult swordfish, tunas, and other tuna species (Cramer and Scott, 1997), today the fleet is made up of domestic vessels (defined as vessels with at least 50% U.S. ownership). The numbers of tunas (bluefin, yellowfin, and blackfin), blue marlin, white marlin, wahoo, and dolphin taken during each set along with the set date are provided for the period 1986-1999. Since this fishery primarily targets tunas, the spatial pattern of fishing effort varies among months (Fig. 2). Some level of fishing effort occurs during each month in most of the region of interest for the present study. For this reason, the presence of tunas and other pelagic species is a potentially useful indicator of their distribution within the study area. The monthly changes in CPUE (number of fish per long-line set) were used to estimate the distributional range of each species in the area of interest. All landings data for the period 1986-1999 were sorted by month. Within each month, the CPUE was estimated within each cell of a 10' longitude x 10' latitude grid (approximately 100 nmile^2 or 343 km^2) superimposed on the study area. The values of all cells containing non-zero CPUE estimates were then color-coded and superimposed on the grid.

The NMFS long-line database provided an estimate of the distributions of adults. Determining the distributions of early life history stages of the target species was more problematic. Data on larval and juvenile abundances are sparse and highly restricted in both space and time. While peer-reviewed and non-peer-reviewed literature provided some limited distributional data, the SEAMAP ichthyoplankton database was the most useful source of information on the distribution and abundance of early-life history stages. Data from the SEAMAP spring and fall plankton surveys as well as additional SEAMAP records spanning other times of the year were examined from 1982 to 1996. Data were grouped by month across all available years and assigned to the same 10' x 10' grid as the long-line database. Distributional maps were coded by the presence or absence of larvae and juveniles. Presence or absence was used because it was not always possible to estimate sampling effort from the dataset.

Additional data on selected pelagic gamefish are available from Dugas et al. (1979) who examined the species composition of the Gulf of Mexico fisheries associated with offshore petroleum platforms. The most highly prized species associated with the offshore blue-water recreational fishery captured by troll and drift fishing were wahoo (*Acanthocybium solanderi*), dolphin (*Coryphaena hippurus*), blue marlin (*Makaira nigricans*), white marlin (*Tetrapturus albidus*) and sailfish (*Istiophorus platypterus*). These and other data were superimposed on the same distributional grid used for the long-line and ichthyoplankton datasets.

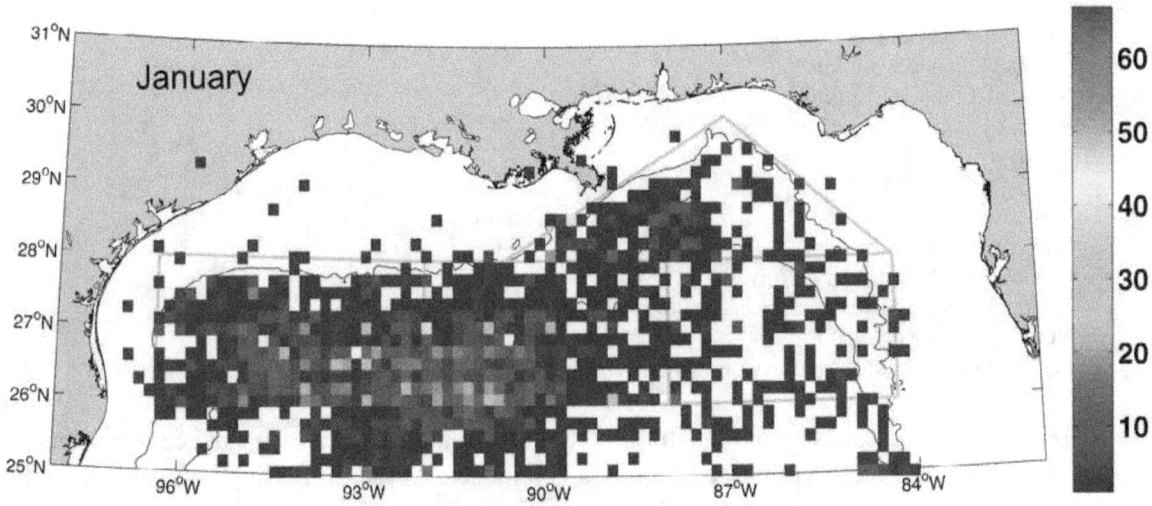

Figure 2.   Commercial long-line fishing effort (number of sets) within 10' x 10' grids (approximately 100 nautical mile²) expressed as mean monthly CPUE (fish per set) from January through December based on data collected from 1987-1999.

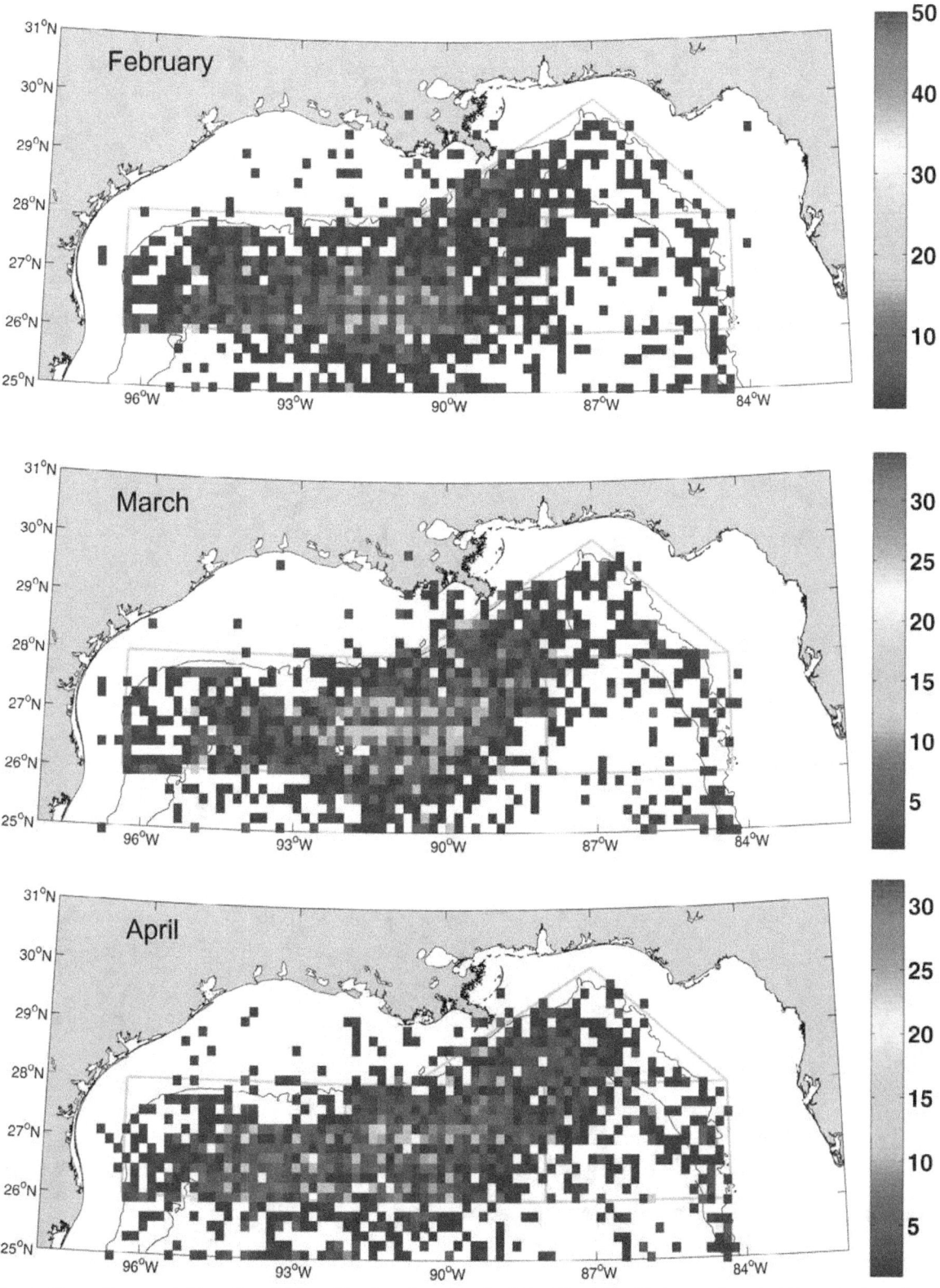

Figure 2 . Commercial long-line fishing effort (number of sets) within 10' x 10' grids (approximately 100 nautical mile^2) expressed as mean monthly CPUE (fish per set) from January through December based on data collected from 1987-1999. (continued)

9

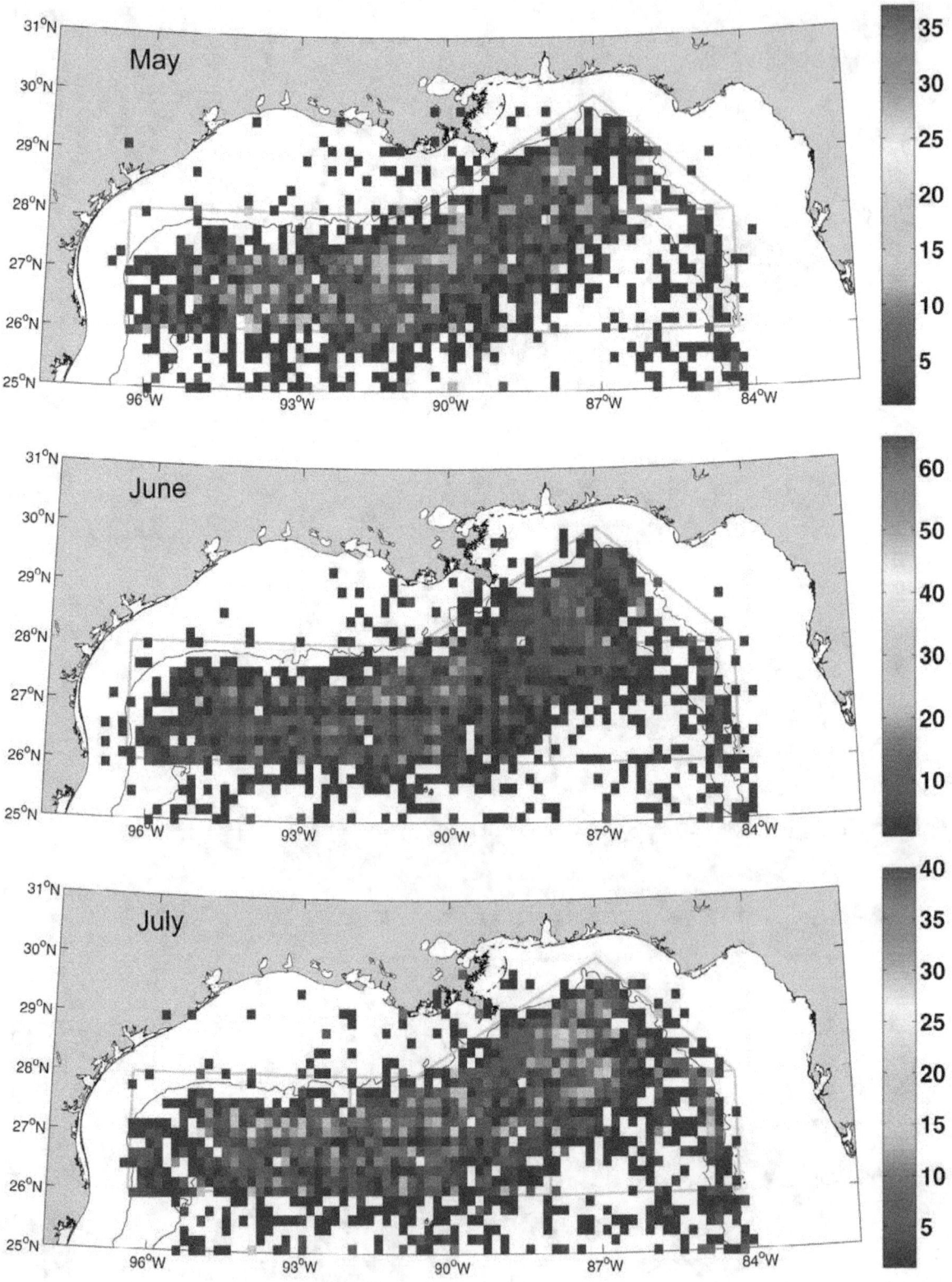

Figure 2. Commercial long-line fishing effort (number of sets) within 10' x 10' grids (approximately 100 nautical mile^2) expressed as mean monthly CPUE (fish per set) from January through December based on data collected from 1987-1999. (continued)

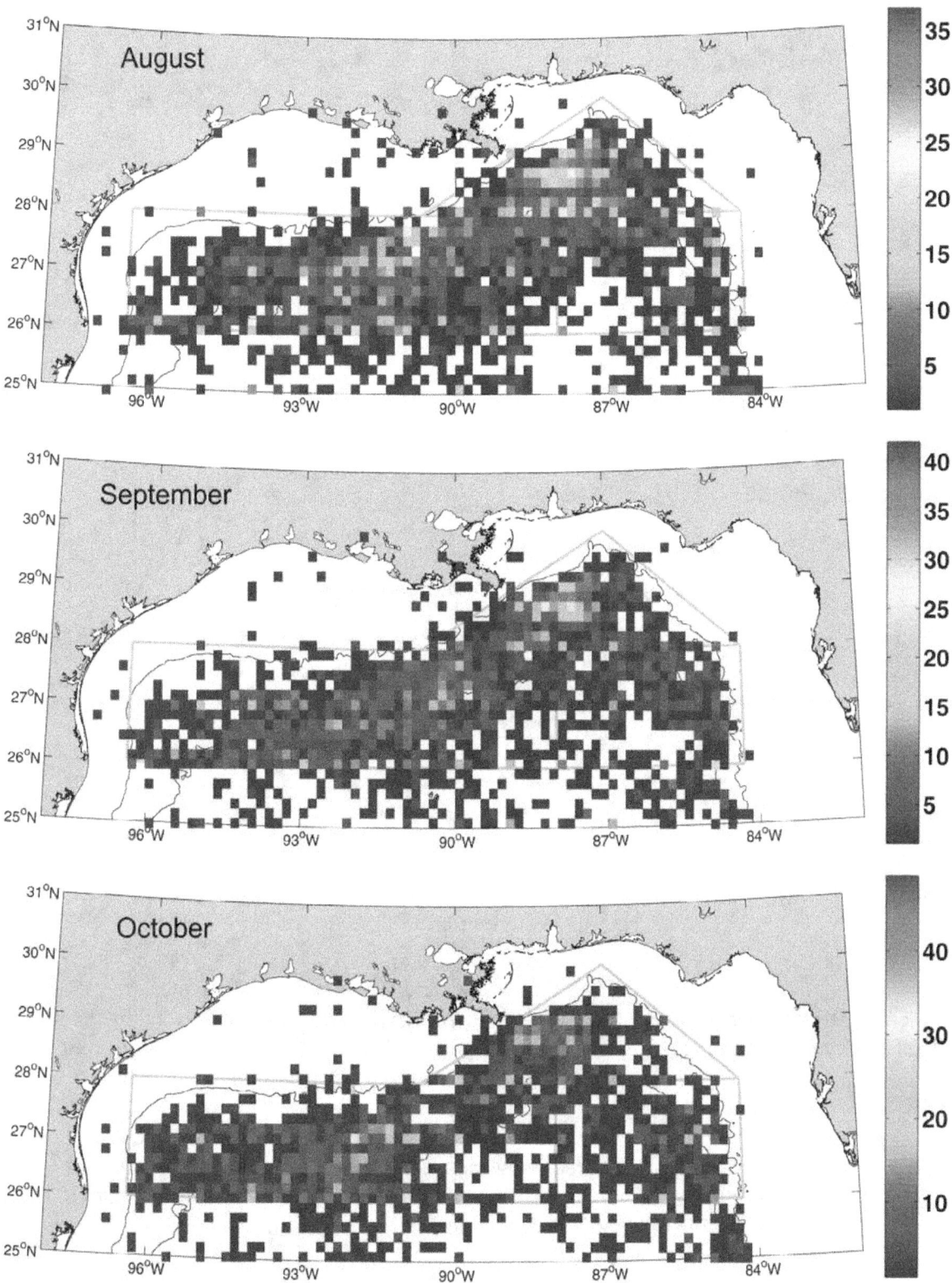

Figure 2. Commercial long-line fishing effort (number of sets) within 10' x 10' grids (approximately 100 nautical mile^2) expressed as mean monthly CPUE (fish per set) from January through December based on data collected from 1987-1999. (continued)

11

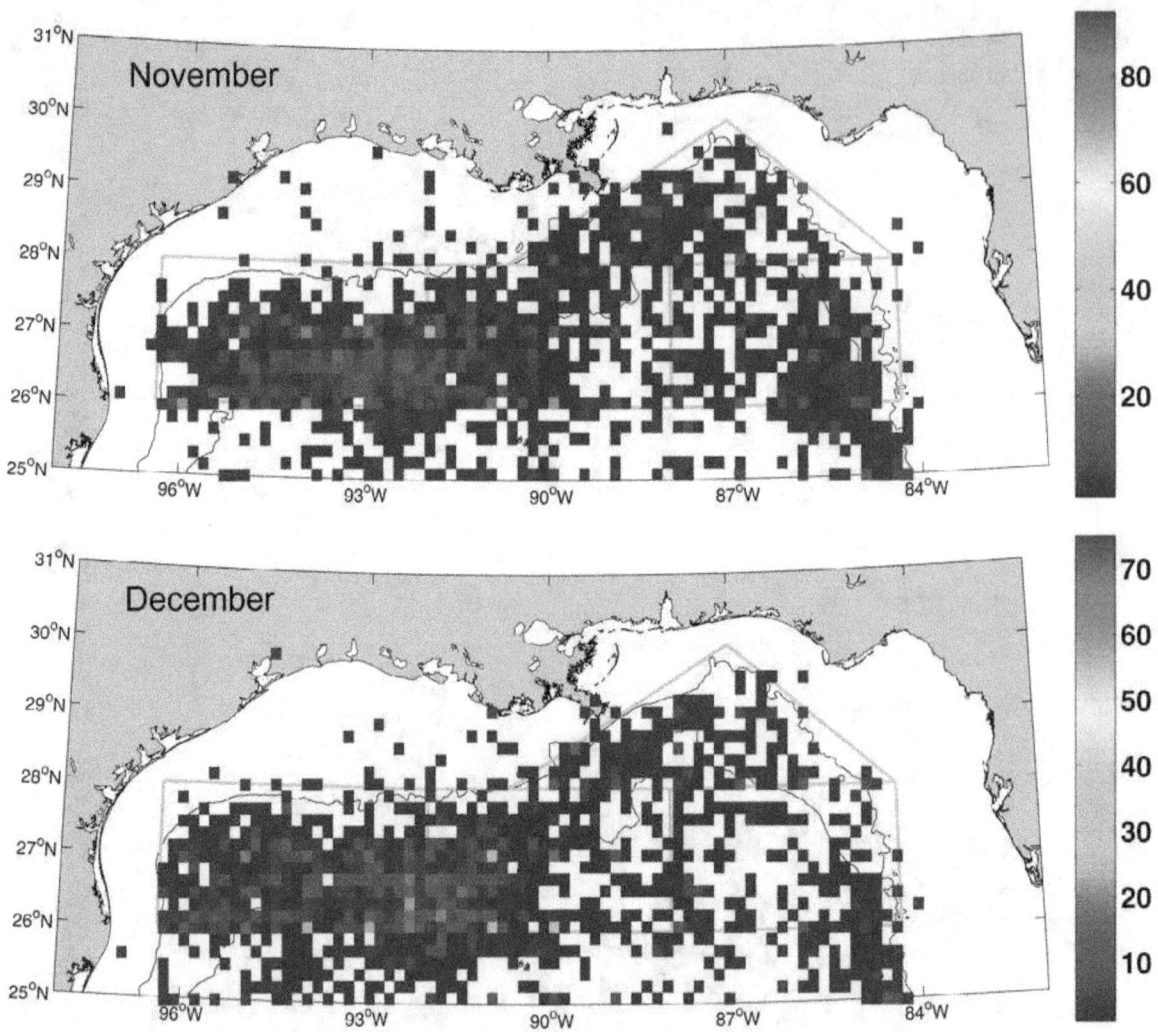

Figure 2. Commercial long-line fishing effort (number of sets) within 10' x 10' grids (approximately 100 nautical mile²) expressed as mean monthly CPUE (fish per set) over the month of December based on data collected from 1987-1999. (continued)

Distributional data were then summarized for adults and larvae (Fig. 3). Juveniles are problematic because gear designed to sample ichthyoplankton is generally avoided by the more capably swimming juveniles. This format differs slightly from that proposed at the inception of this study in-order to predict distributions on a month-by-month rather than annual basis.

- The potential presence of a particular stage in any of the cells within the grid was ranked according to three categories: confirmed, reasonable inference, and unreported. Confirmed presence was assigned when a physical sample of the relevant stage of a particular taxon had been reported in the primary literature as being present within a cell.

- Given the high mobility of most of the adults of species in this study, we assumed that an individual that was detected in any cell of the study grid, could reasonably have traveled a distance of two additional cells around the detection cell within a month of collection.

■ Reasonable inference for adults stages was therefore assigned to any cells within which, there was no confirmed presence, providing the cells were located within a radius of two cell distances (up to approximately 37 km) of a cell with a confirmed presence. This is probably a very conservative estimate of the distances that some of these fish can travel. Reasonable inference was also assigned to any cells that were bounded by four or more cells also designated with the reasonable inference category. Finally, any regions within the study area that were completely surrounded by cells designated as confirmed or reasonable inference, were also assigned reasonable inference.

■ Reasonable inference was also assigned to any cells that fell within a region where the distribution had been reported in a document that synthesized the results from other datasets. For example, the National Ocean Service (NOS) Strategic Assessment Data Atlas (NOS, 1985) contains distributional maps for several of the target species derived from analysis of other studies. Such maps were digitized and scaled to our study area.

■ Finally, all cells that did not fall into the confirmed or reasonable were assigned an unreported category.

For larvae, reasonable inference was confined to cells from which no physical sample had been reported, which were contiguous with confirmed cells.

■ Most larvae are present when the surface waters of the Gulf of Mexico are warm. Thus growth rates are rapid and the larval duration probably does not exceed 14 days.

■ Assuming that larvae are drifting in slow currents (0.1 knots or less), net advection during a four week period should not exceed 33 nautical miles and would likely be considerably less. Therefore, we applied a rather conservative distance and assigned the reasonable inference category to any cells within which, larvae had not been detected, but which were within one cell of a confirmed cell, or were entirely enclosed by cells designated as confirmed or reasonable inferences.

■ As with the adult distributions, all cells that did not fall into the confirmed or reasonable were assigned an unreported category.

Estimation of spatial distributions based on these decision rules would only work for regions within the study area where sampling effort was sufficiently dense that there would not be any empty zones after the application of our classification strategy (Fig. 3). We evaluated the spatial coverage of the NMFS and SEAMAP datasets by assuming that a confirmed sample was collected at each recorded location during each month and then applied our classification strategy. For adult fishes, the results indicated that there were no gaps in our predicted distributions throughout the study area during all months (Fig. 4).

13

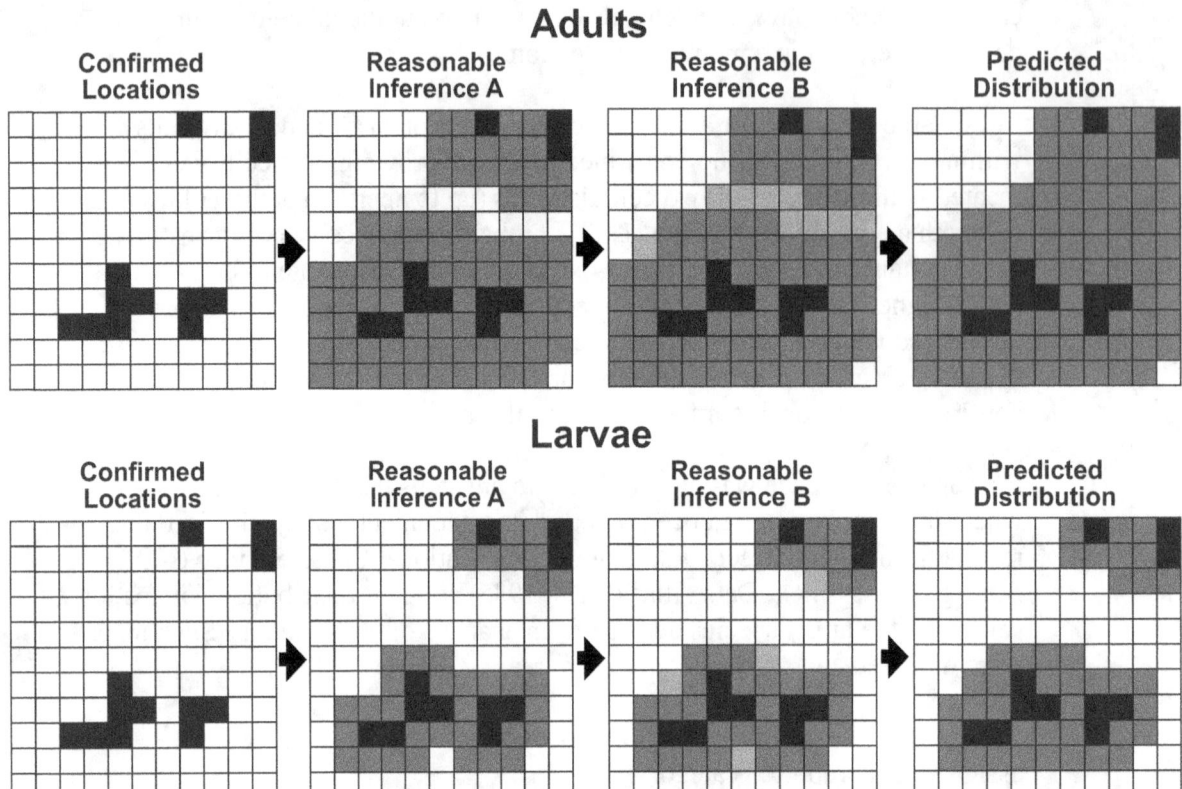

Figure 3. Strategy for classifying distributions of adult fishes (top) and larval/juvenile fishes (bottom). For adults, all cells within 2 grid squares of each confirmed location (■) were assigned the reasonable inference category (■). Next, any empty cells bounded by four or more reasonable inference cells (■) were also assigned the reasonable inference category. For larval/juvenile fishes, all cells within 1 grid square of each confirmed location were assigned the reasonable inference category (identified as reasonable inference A), and then any empty cells bounded by four or more cells classified as reasonable inference, were also assigned the reasonable inference category (identified as reasonable inference B). After these rules had been applied, any pockets of the map that were completely surrounded by either confirmed locations or reasonable inferences, were assigned the category of reasonable inference.

Larval and juvenile distributions based on SEAMAP samples were more problematic. These surveys are primarily designed to quantify distributions over the shelf rather than slope water. After application of our classification strategy, with the assumption of detection of at least one individual at every sampling location, there were still large areas of the study zone where there was insufficient coverage to infer larval and juvenile distributions (Fig. 5). During January, most of the eastern zone and the eastern half of the northern zone were not covered. In February, coverage was sparse in all zones, but was particularly low in the northern and eastern zones. During March, there was no coverage in any of the study area zones (Fig. 5). Coverage improved during April, May, and June although during these months, each zone contained areas where there was no predictive capability (Fig. 5). In July, predictive coverage was generally confined to the northern and western periphery of the western and northern zones. This coverage expanded in August and September to include the eastern peripheries of the northern and eastern zones, however, the western, central and eastern zones were generally poorly covered during summer (Fig. 5). During October and November, predictive coverage was extremely limited. December provided good coverage of the northern zone, and the northern halves of the central and eastern zones (Fig. 5). In spite of the gaps that limit the utility of the SEAMAP dataset to predict the distributions of larvae and juveniles in the deepwater zones of this review, most zones were well covered during April-June when the majority of species are spawning in the Gulf of Mexico.

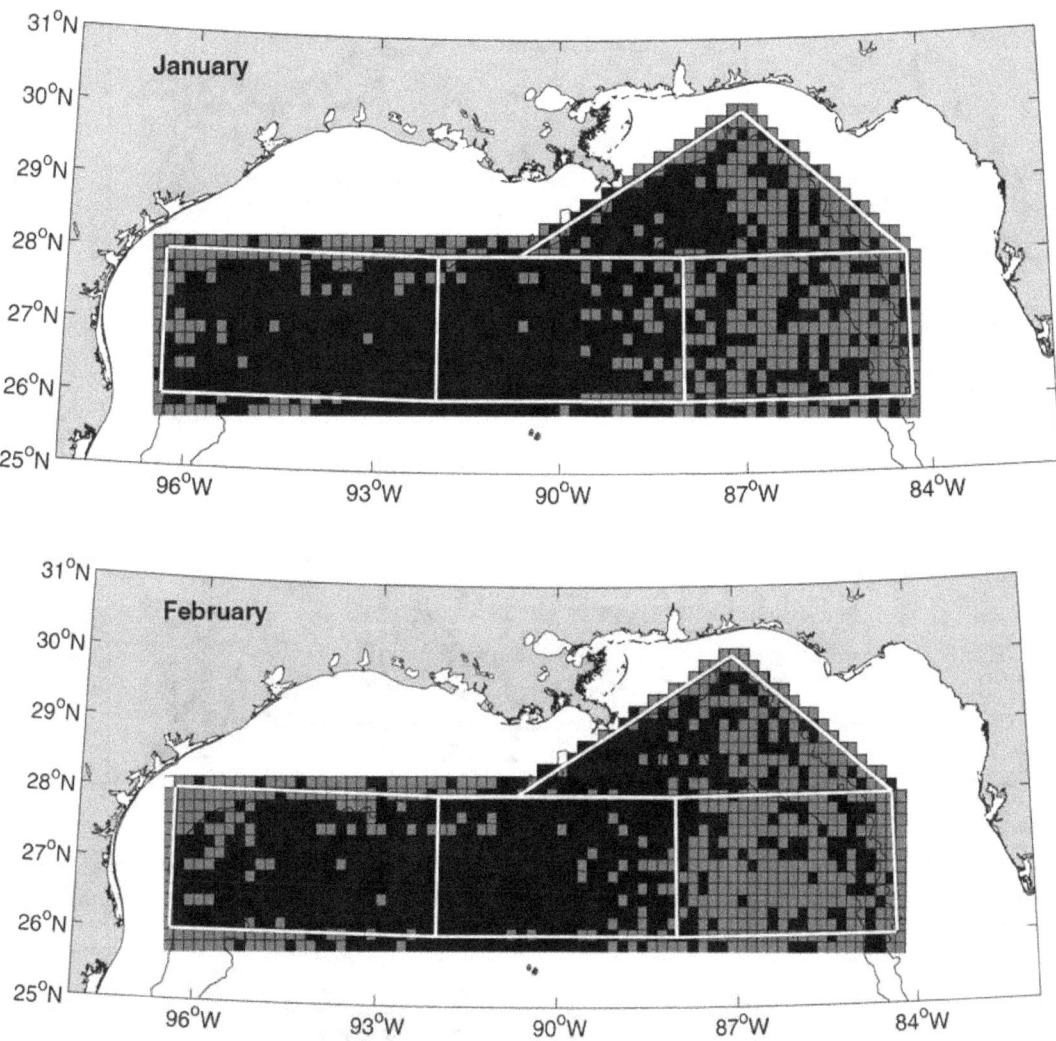

Figure 4. Predictive coverage by the NMFS long-line database of the study area from January through December with the spatial coverage of predicted distributions of adult fishes obtained using the decision rules outlined in Fig. 3. The decision rules assumed that one fish was detected in each of the grid cell containing a longline record. The presence of individuals in each grid cell is coded as: confirmed (■), reasonable inference (▨) or unreported (□).

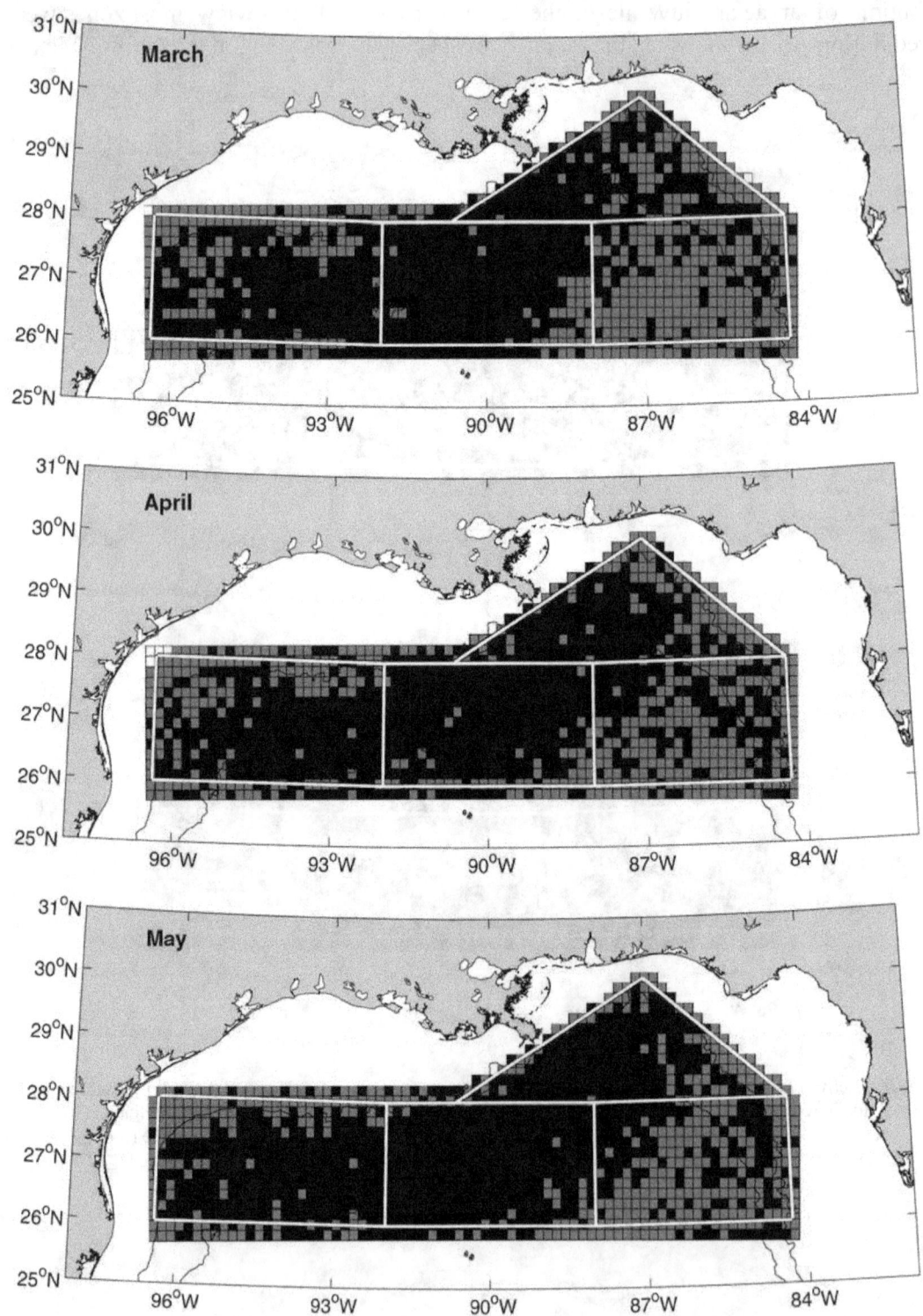

Figure 4. Predictive coverage by the NMFS longline database of the study area from January through December with the spatial coverage of predicted distributions of adult fishes obtained using the decision rules outlined in Fig. 3. The decision rules assumed that one fish was detected in each of the grid cell containing a longline record. The presence of individuals in each grid cell is coded as: confirmed (■), reasonable inference (▨) or unreported (□). (continued)

16

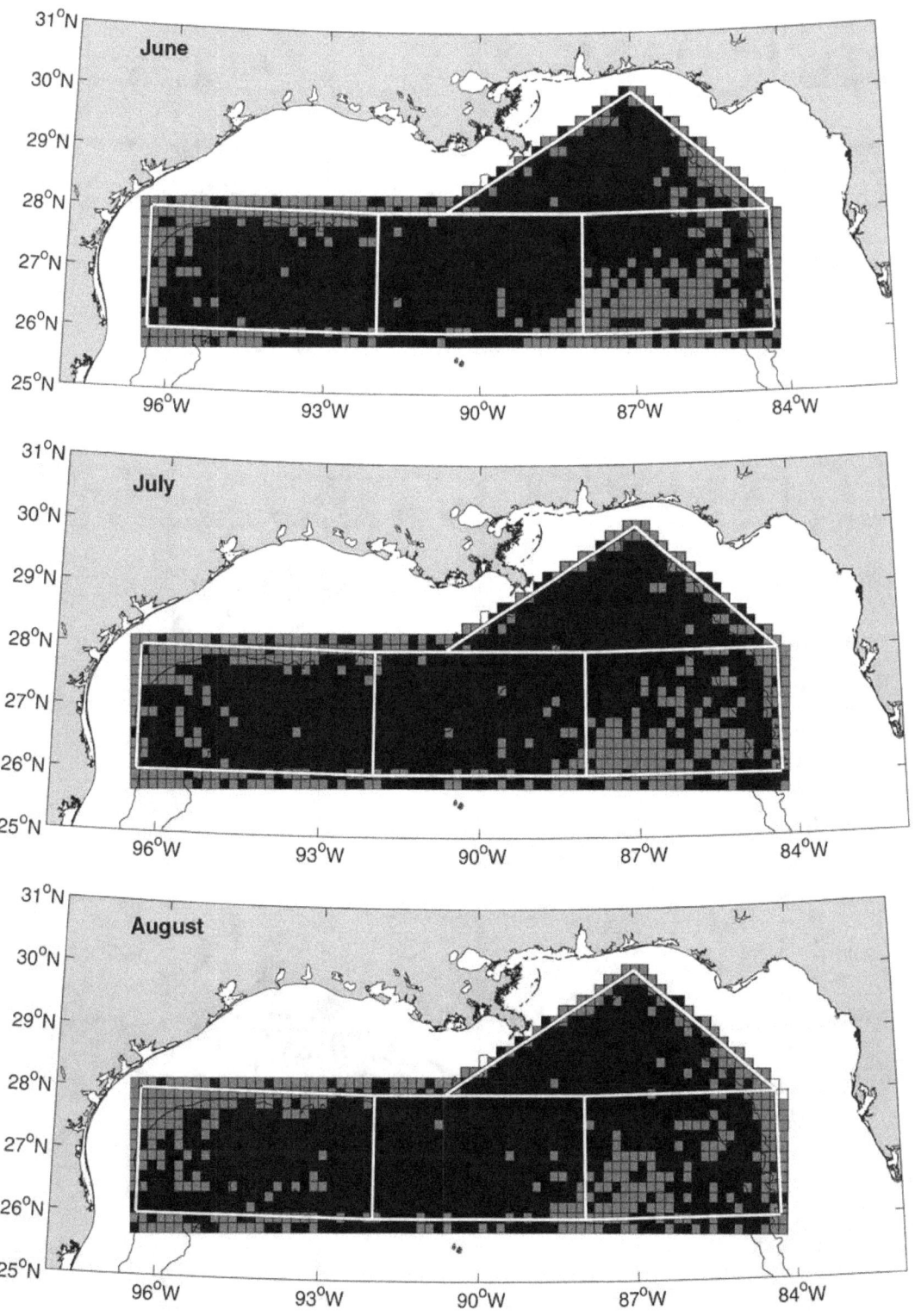

Figure 4. Predictive coverage by the NMFS longline database of the study area from January through December with the spatial coverage of predicted distributions of adult fishes obtained using the decision rules outlined in Fig. 3. The decision rules assumed that one fish was detected in each of the grid cell containing a longline record. The presence of individuals in each grid cell is coded as: confirmed (■), reasonable inference (▨) or unreported (□). (continued)

17

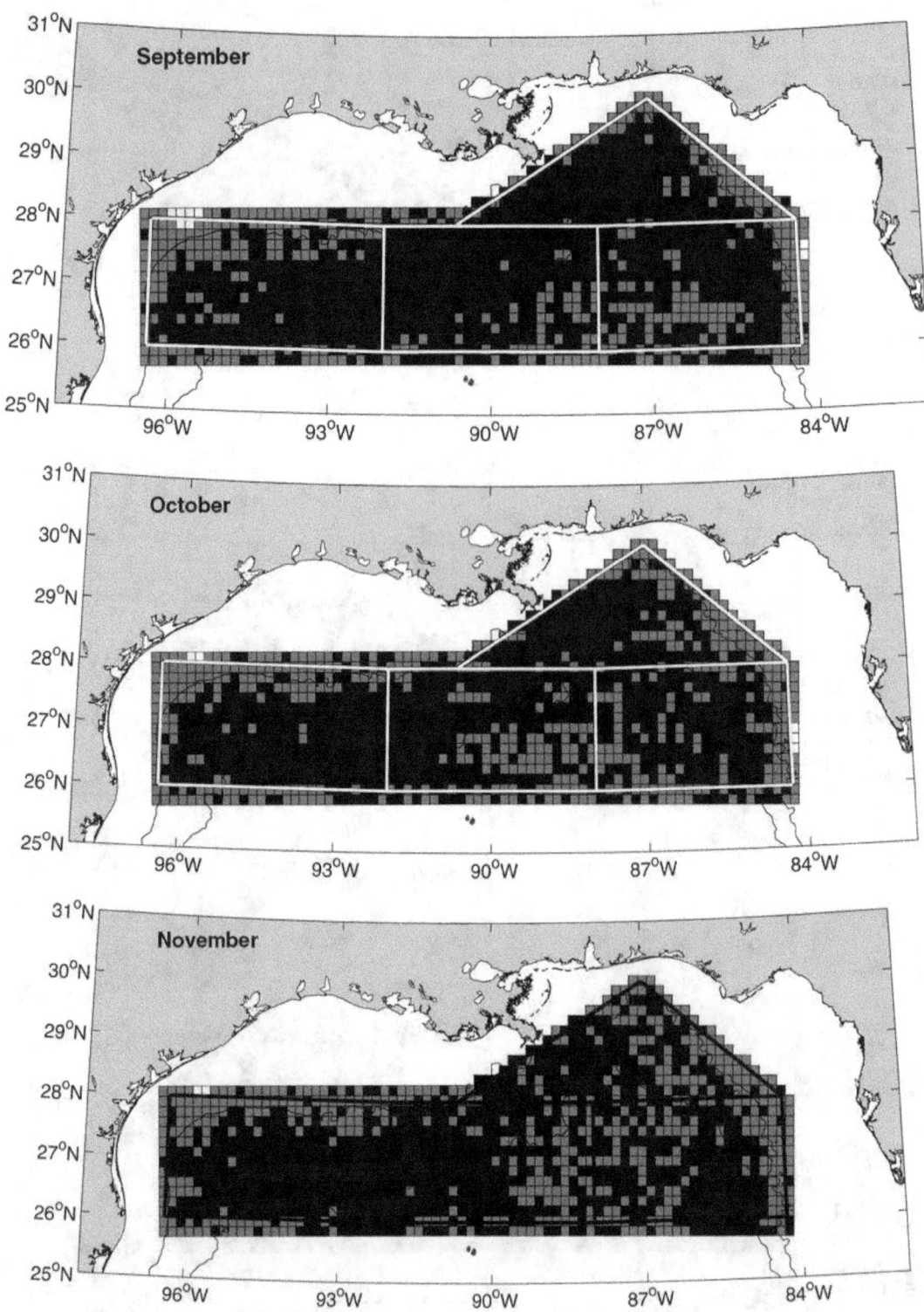

Figure 4. Predictive coverage by the NMFS longline database of the study area from January through December with the spatial coverage of predicted distributions of adult fishes obtained using the decision rules outlined in Fig. 3. The decision rules assumed that one fish was detected in each of the grid cell containing a longline record. The presence of individuals in each grid cell is coded as: confirmed (■), reasonable inference (▦) or unreported (□). (continued)

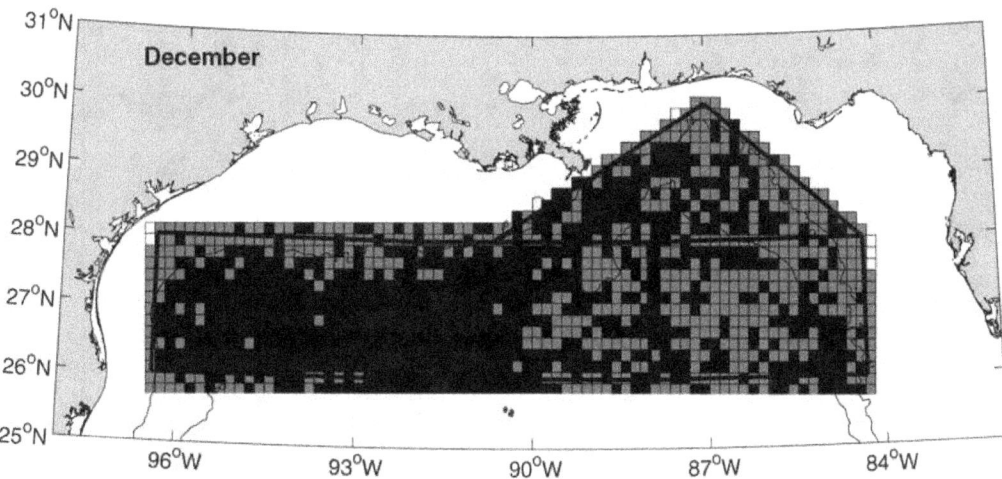

Figure 4. Predictive coverage by the NMFS longline database of the study area from January through December with the spatial coverage of predicted distributions of adult fishes obtained using the decision rules outlined in Fig. 3. The decision rules assumed that one fish was detected in each of the grid cell containing a longline record. The presence of individuals in each grid cell is coded as: confirmed (■), reasonable inference (▓) or unreported (□). (continued)

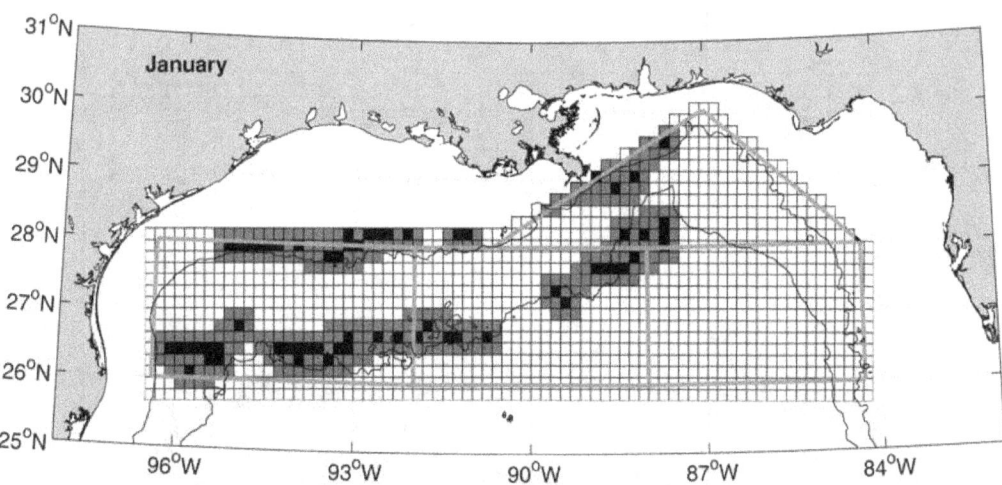

Figure 5. Predictive coverage by the SEAMAP database of the study area from January through December with the spatial coverage of predicted distributions of larval and juvenile fishes using the decision rules outlined in Fig. 3. The decision rules assumed that one fish was detected in each of the grid cell containing a SEAMAP sample. The presence of individuals in each grid cell is coded as: confirmed (■), reasonable inference (▓) or unreported (□). (continued)

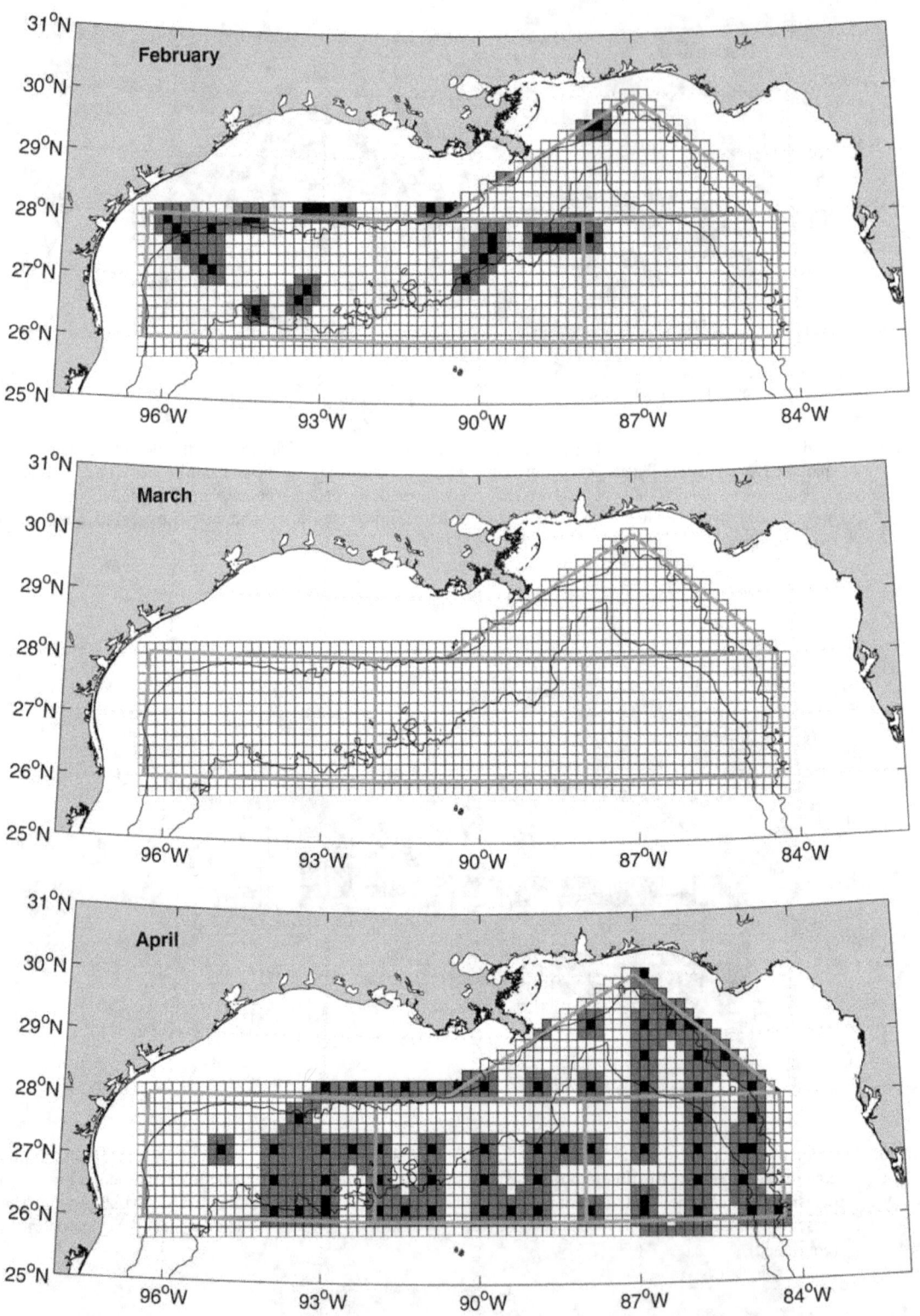

Figure 5. Predictive coverage by the SEAMAP database of the study area from January through December with the spatial coverage of predicted distributions of larval and juvenile fishes using the decision rules outlined in Fig. 3. The decision rules assumed that one fish was detected in each of the grid cell containing a SEAMAP sample. The presence of individuals in each grid cell is coded as: confirmed (■), reasonable inference (▨) or unreported (□). (continued)

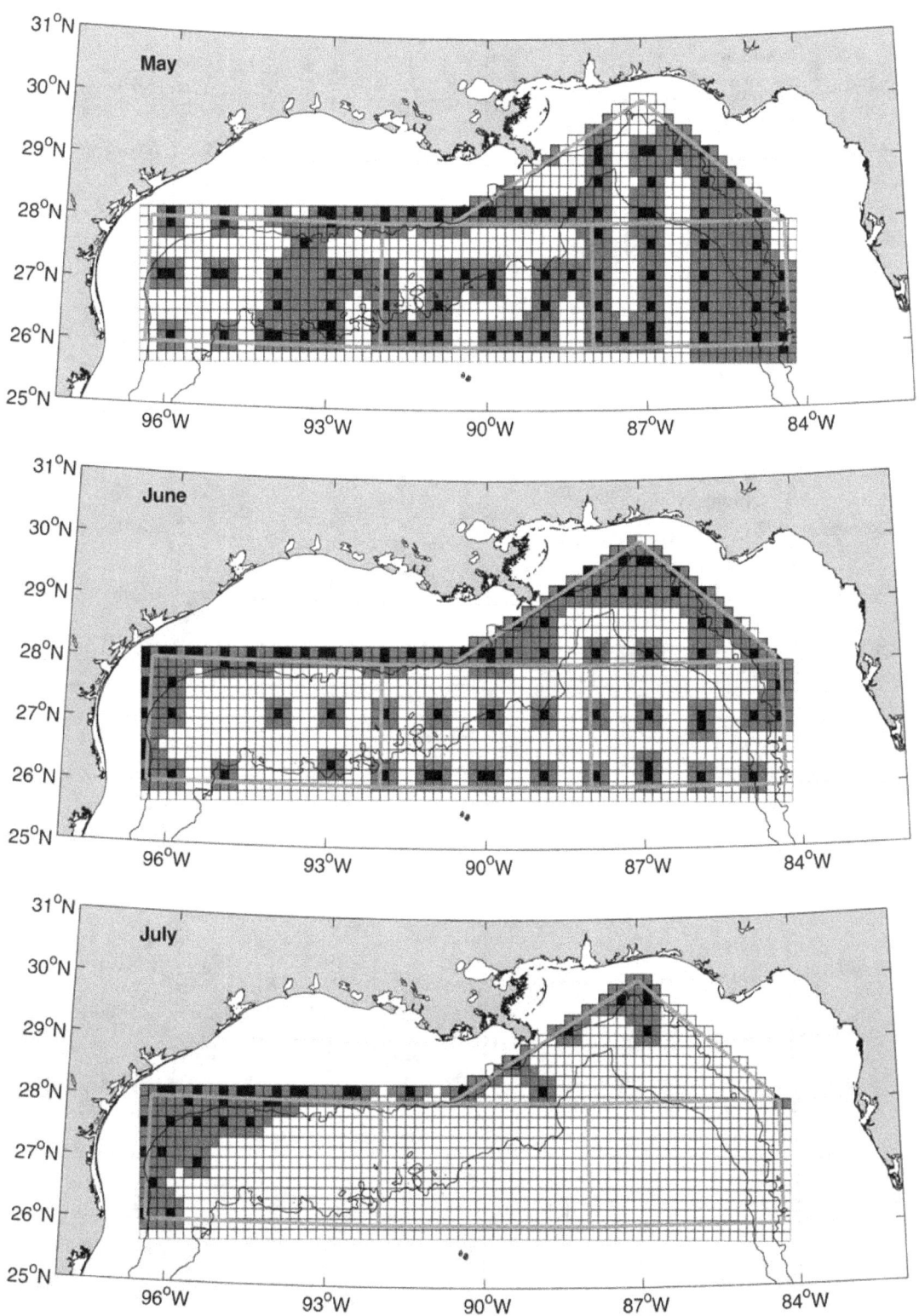

Figure 5. Predictive coverage by the SEAMAP database of the study area from January through December with the spatial coverage of predicted distributions of larval and juvenile fishes using the decision rules outlined in Fig. 3. The decision rules assumed that one fish was detected in each of the grid cell containing a SEAMAP sample. The presence of individuals in each grid cell is coded as: confirmed (■), reasonable inference (▨) or unreported (□). (continued)

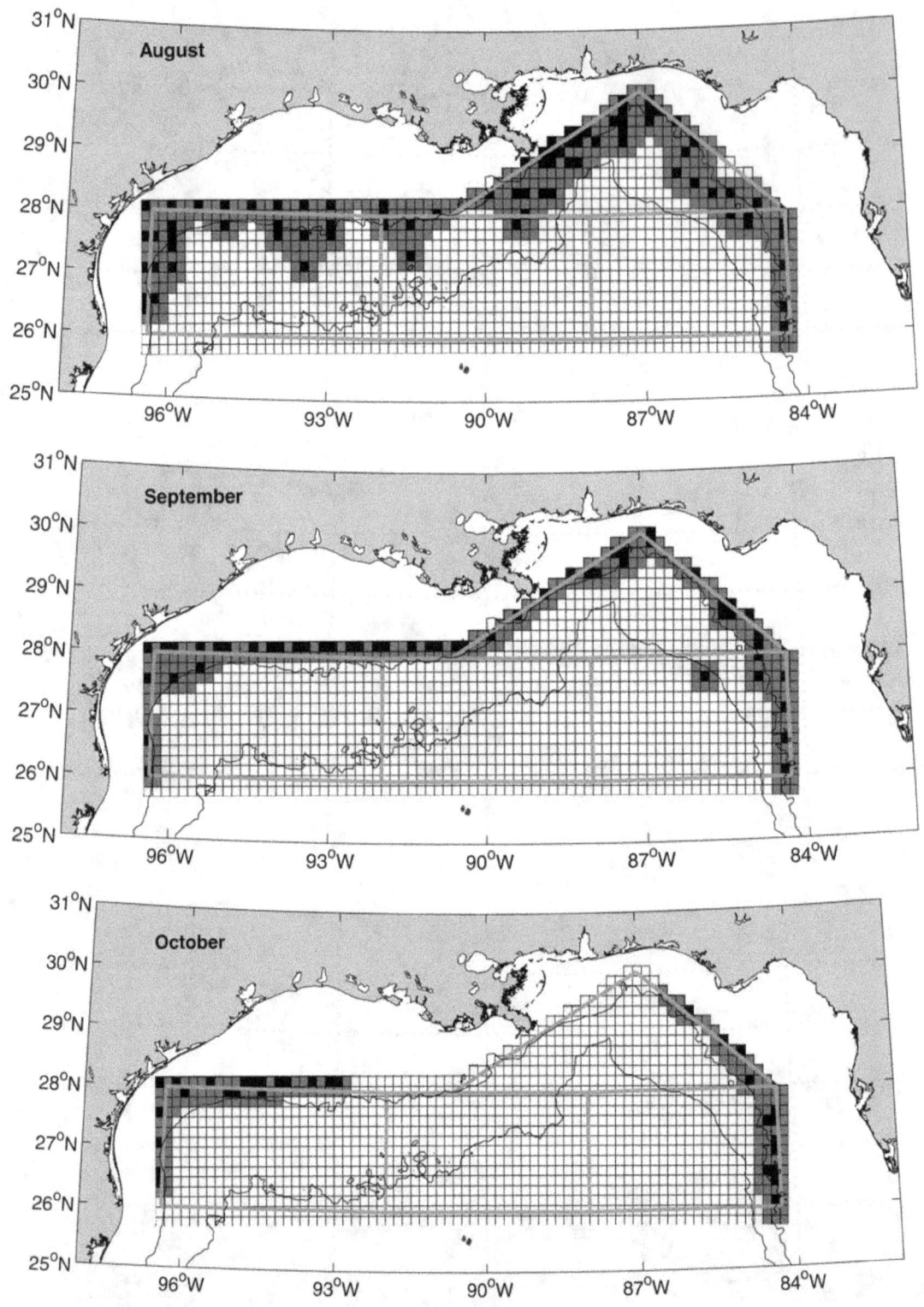

Figure 5. Predictive coverage by the SEAMAP database of the study area from January through December with the spatial coverage of predicted distributions of larval and juvenile fishes using the decision rules outlined in Fig. 3. The decision rules assumed that one fish was detected in each of the grid cell containing a SEAMAP sample. The presence of individuals in each grid cell is coded as: confirmed (■), reasonable inference (▨) or unreported (□). (continued)

22

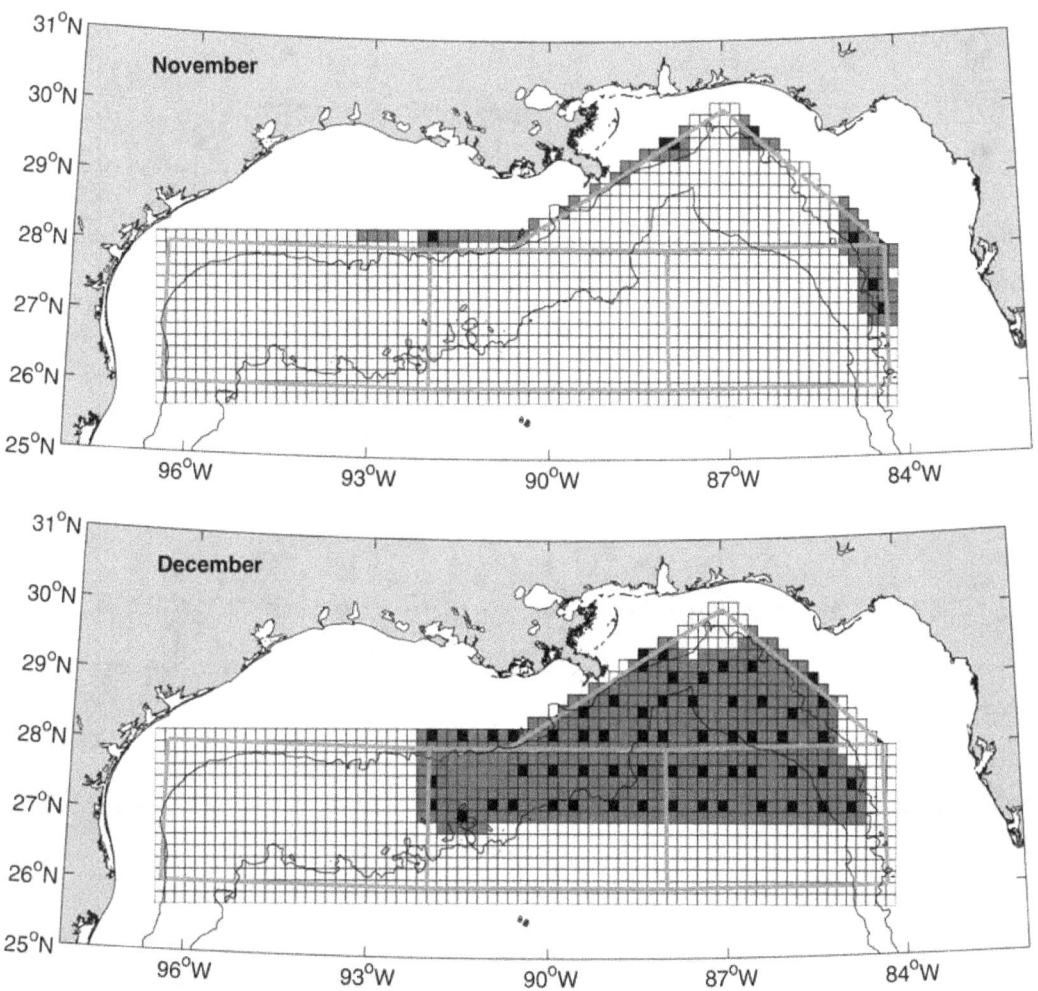

Figure 5. Predictive coverage by the SEAMAP database of the study area from January through December with the spatial coverage of predicted distributions of larval and juvenile fishes using the decision rules outlined in Fig. 3. The decision rules assumed that one fish was detected in each of the grid cell containing a SEAMAP sample. The presence of individuals in each grid cell is coded as: confirmed (■), reasonable inference (▨) or unreported (□). (continued)

4   Life History and Distributional Summaries
4.1   Yellowfin Tuna (*Thunnus albacares*)

Yellowfin tuna are found throughout the tropical oceans of the world and are abundant in the Gulf of Mexico, particularly during the spring and summer (Southeast Fisheries Science Center, 1992). Yellowfin tuna constitute one of the top ten most important species harvested commercially in the Gulf based on both landing tonnage and value (Adams, 1996). The distribution of adults of these migratory pelagic fish is believed to be determined by the presence of prey (small pelagic fishes and squids) and water temperature (Southeast Fisheries Science Center, 1992).

4.1.1   Adult Distributions

It has been suggested that tunas are more abundant near frontal regions (e.g., Maul et al., 1984) however, Power and May (1991) found no such association between the yellowfin tuna commercial long-line CPUE and sea surface temperature in the northwestern Gulf of Mexico. NMFS long-line data indicated that adult yellowfin tuna were abundant within most of the study areas during all months of the year (Fig. 6). Lowest abundances were noted in the eastern zone and the eastern half of the northern zone from Jan-May (Fig. 6). NOS (1985) suggest a broad distribution that encompasses the entire region of this study (Fig. 7) although monthly distributional data were not available.

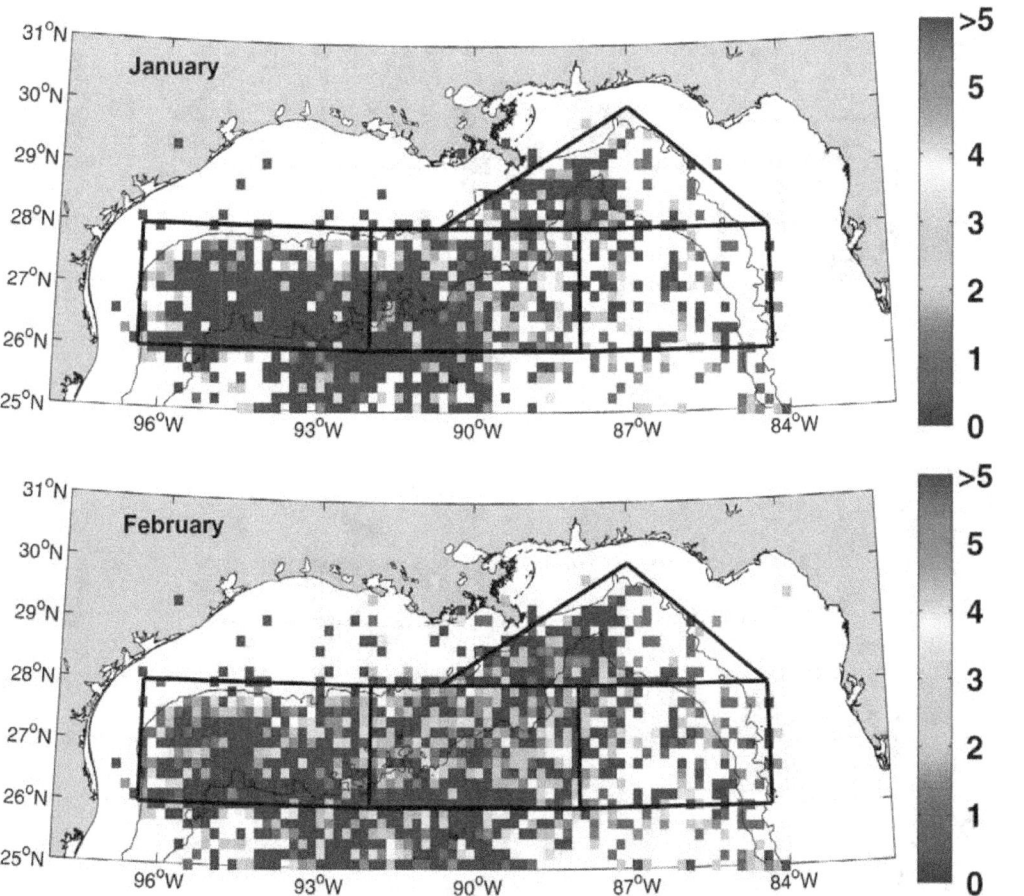

Figure 6.   Catch per unit effort of adult yellowfin tuna from the commercial long-line fishery. Each square represents the mean catch-per-unit-effort (tuna per set) taken within a 10' x 10' region for the month of January over the period 1986-1999. The maximum CPUEs were: January =123; February = 210; March = 113; April = 121; May = 105; June = 178; July = 158; August = 140; September = 186; October = 158; November = 260; December = 140.)

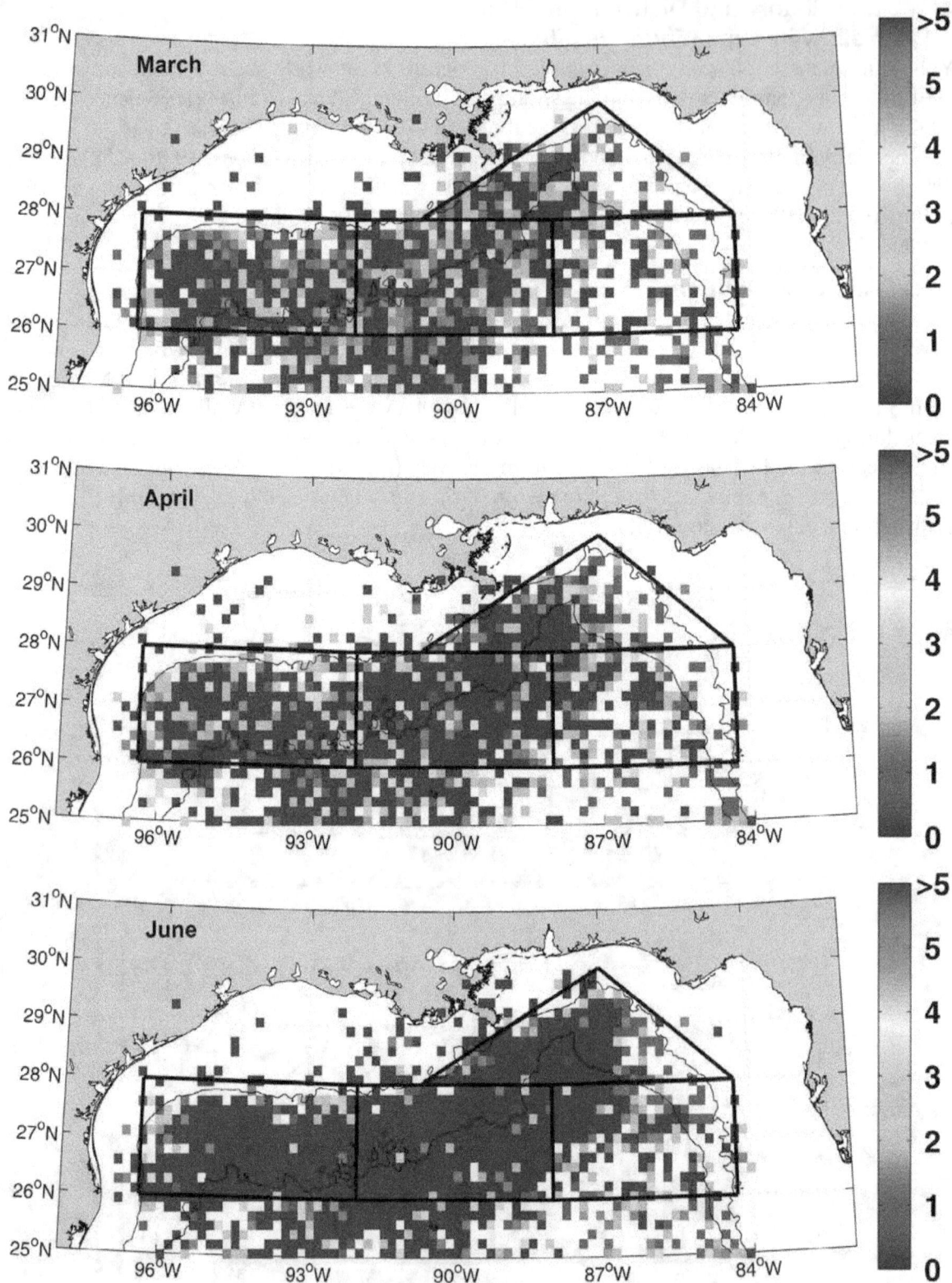

Figure 6.   Catch per unit effort of adult yellowfin tuna from the commercial long-line fishery. Each square represents the mean catch-per-unit-effort (tuna per set) taken within a 10' x 10' region for the month of January over the period 1986-1999. The maximum CPUEs were: January =123; February = 210; March = 113; April = 121; May = 105; June = 178; July = 158; August = 140; September = 186; October = 158; November = 260; December = 140. (continued)

26

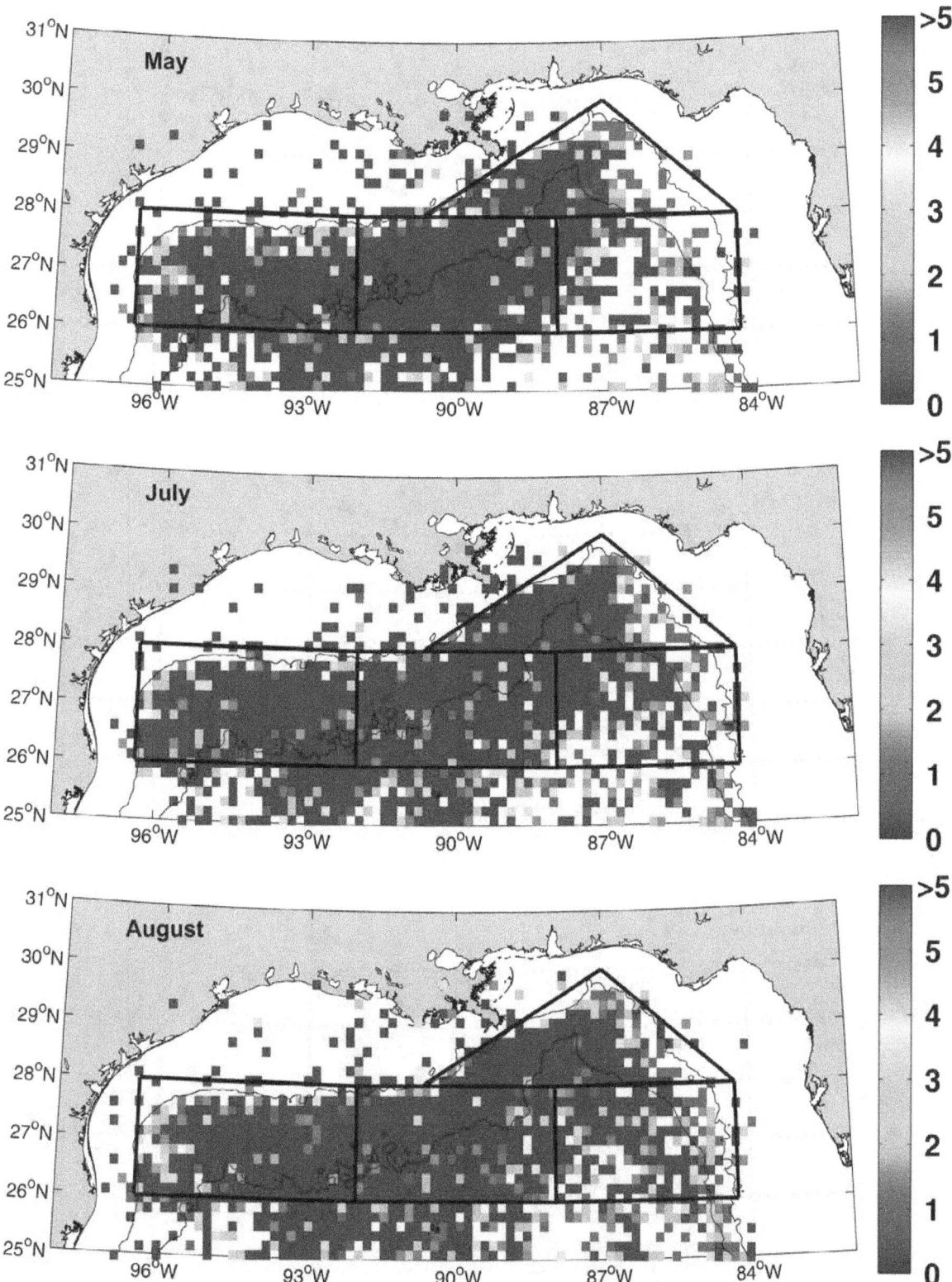

Figure 6.   Catch per unit effort of adult yellowfin tuna from the commercial long-line fishery. Each square represents the mean catch-per-unit-effort (tuna per set) taken within a 10' x 10' region for the month of January over the period 1986-1999. The maximum CPUEs were: January =123; February = 210; March = 113; April = 121; May = 105; June = 178; July = 158; August = 140; September = 186; October = 158; November = 260; December = 140. (continued)

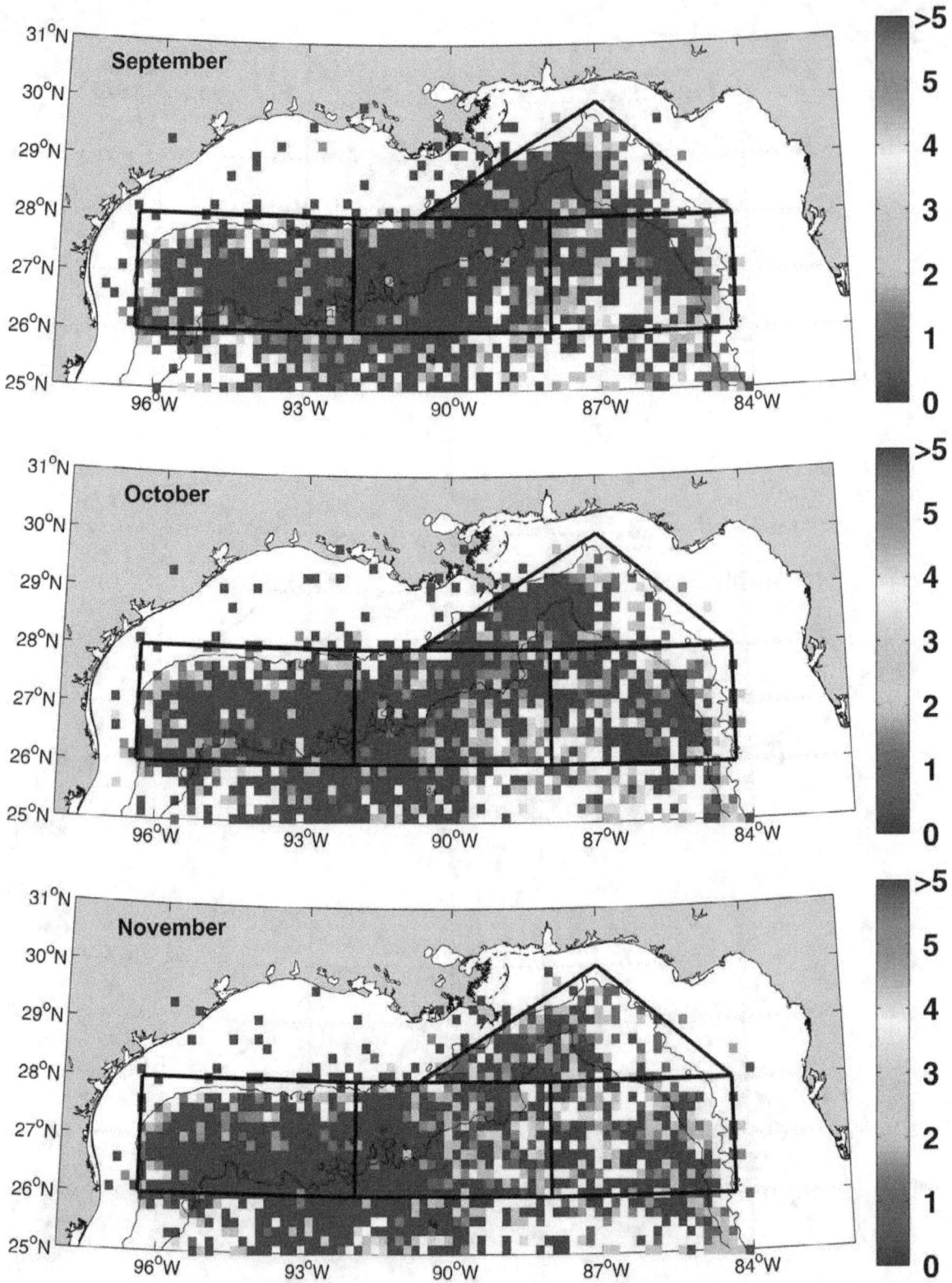

Figure 6. Catch per unit effort of adult yellowfin tuna from the commercial long-line fishery. Each square represents the mean catch-per-unit-effort (tuna per set) taken within a 10' x 10' region for the month of January over the period 1986-1999. The maximum CPUEs were: January =123; February = 210; March = 113; April = 121; May = 105; June = 178; July = 158; August = 140; September = 186; October = 158; November = 260; December = 140. (continued)

28

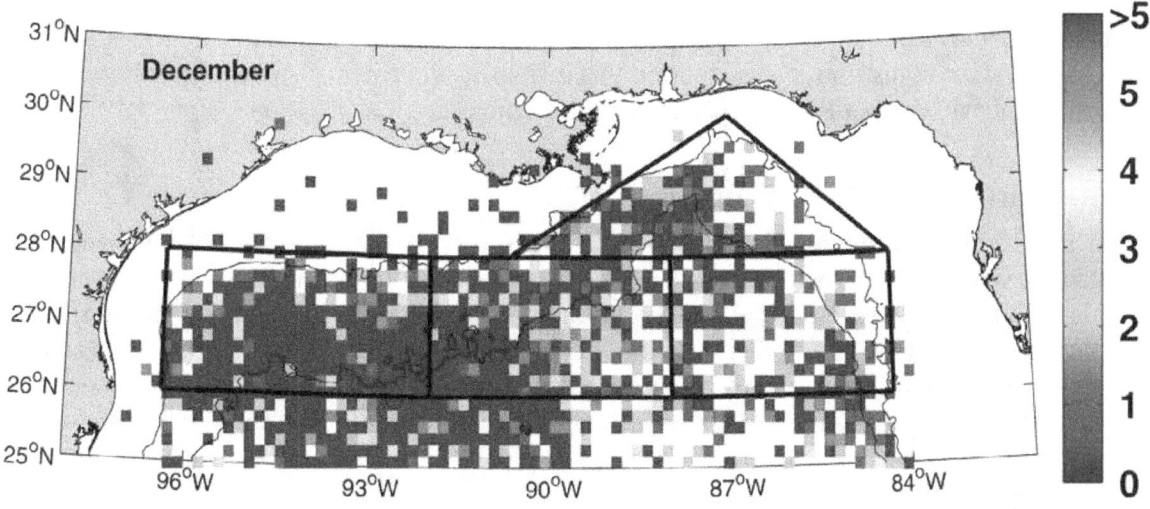

Figure 6. Catch per unit effort of adult yellowfin tuna from the commercial long-line fishery. Each square represents the mean catch-per-unit-effort (tuna per set) taken within a 10' x 10' region for the month of January over the period 1986-1999. The maximum CPUEs were: January =123; February = 210; March = 113; April = 121; May = 105; June = 178; July = 158; August = 140; September = 186; October = 158; November = 260; December = 140. (continued)

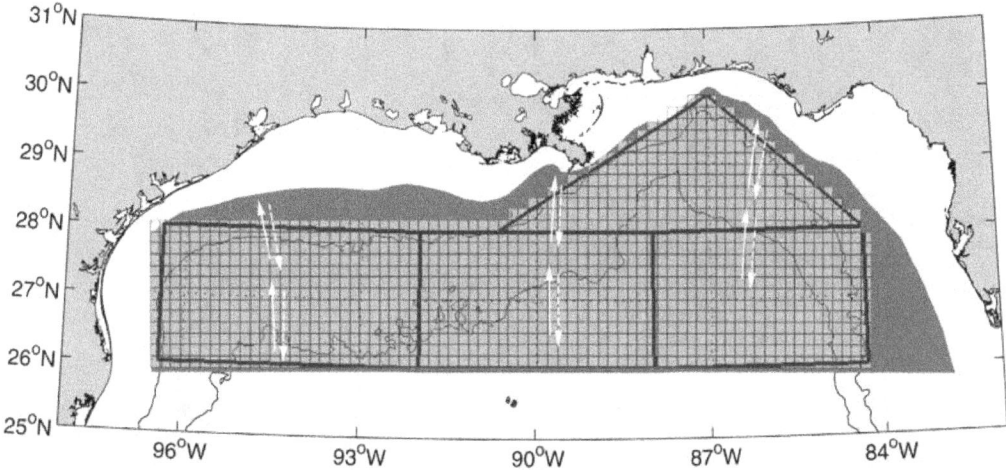

Figure 7. Distribution of adult yellowfin tuna from NOS (1985). Note that the offshore distribution has been cut-off south of the southern limit or the present study zone. The solid arrows indicate the summer migration inshore and the dashed arrows indicate the winter migration offshore. No breakdown of distributions by months was presented.

4.1.2 Reproduction

Spawning occurs from April through June in the western Atlantic and Gulf of Mexico (Southeast Fisheries Science Center, 1992). The presence of larvae in the northern Gulf from April through September with a peak during May, June and July (Ditty et al. 1988) suggests that spawning may extend throughout summer. Examinations of ovaries and testes from yellowfin tuna collected in the Gulf from August through February did not reveal evidence of gonadal development conducive for spawning (Goldberg and Herring-Dyal, 1981). A subsequent study conducted by Grimes and Lang (1992) suggested that reproduction extends through August and September. They surveyed the waters off the Mississippi plume for yellowfin tuna larvae and hypothesized that reproduction extends from mid-summer through September. This result was supported by Lang et al. (1994) based on back-calculated birth dates of larvae collected off the Mississippi River plume. Larval distributions suggested that a significant center of spawning activity might occur near the Mississippi River discharge plume (Grimes and Lang, 1992).

4.1.3   Larval/Juvenile Distributions

Tunas produce large numbers of small eggs that hatch within a short period (1-2 days) and the resultant larvae grow rapidly (Klawe and Shimada, 1959). Distinguishing the larvae of yellowfin and blackfin tunas is problematic (Grimes and Lang, 1992; Lang et al. 1994) and this can complicate the interpretation of their distributional patterns based on collections of early stage larvae. In Grimes and Lang's survey of larval distributions off the Mississippi, larvae were concentrated off the Mississippi discharge plume (Fig. 8) in intermediate salinity surface waters on the Gulf of Mexico side of the riverine frontal region. Studies by Lang et al. (1994) during July and September also indicated high catches in frontal regions at intermediate salinities (~31 psu) and temperatures (29.8 °C). Scarcity of definitively identified *T. albacares* larvae and juveniles in the SEAMAP dataset (Fig. 9) makes interpretation of their seasonal distribution in the study area difficult.

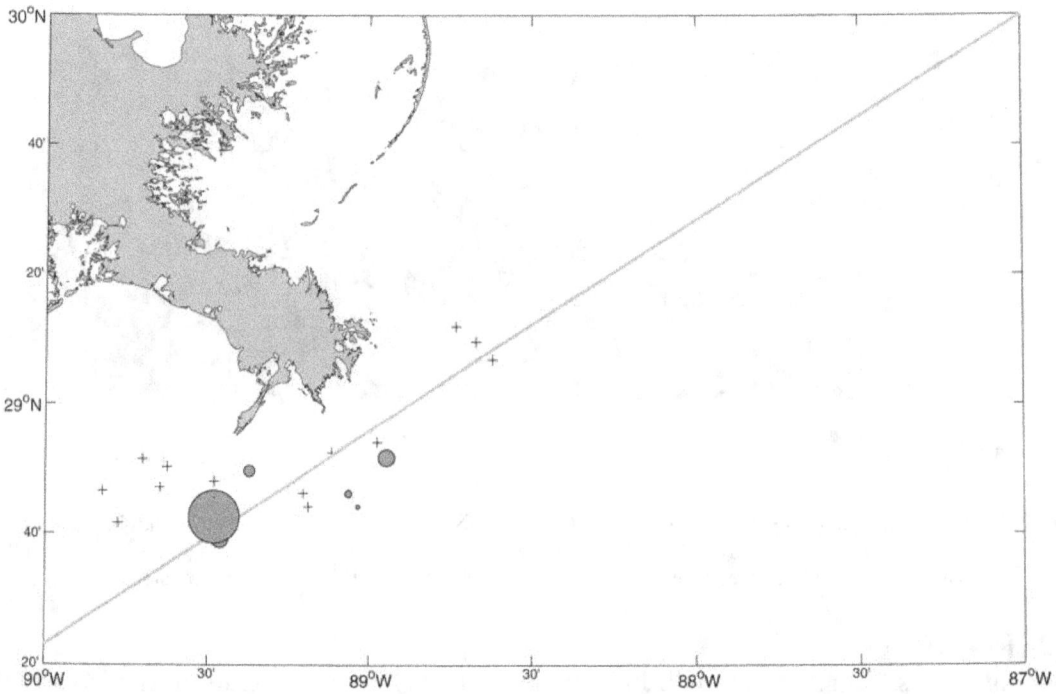

Figure 8.   Distribution and abundance of yellowfin tuna larvae off the Mississippi River plume during early September, 1987. Crosses (+) indicate larvae present at densities of 1-10 100 m^{-2} and the diameters of the largest circle indicates 301 larvae 100 m^{-2}. The gray line indicates the western boundary of the northern zone of this study. Data from Kathy Lang, NMFS, Oregon II Cruise 169, September 3-6, 1987.

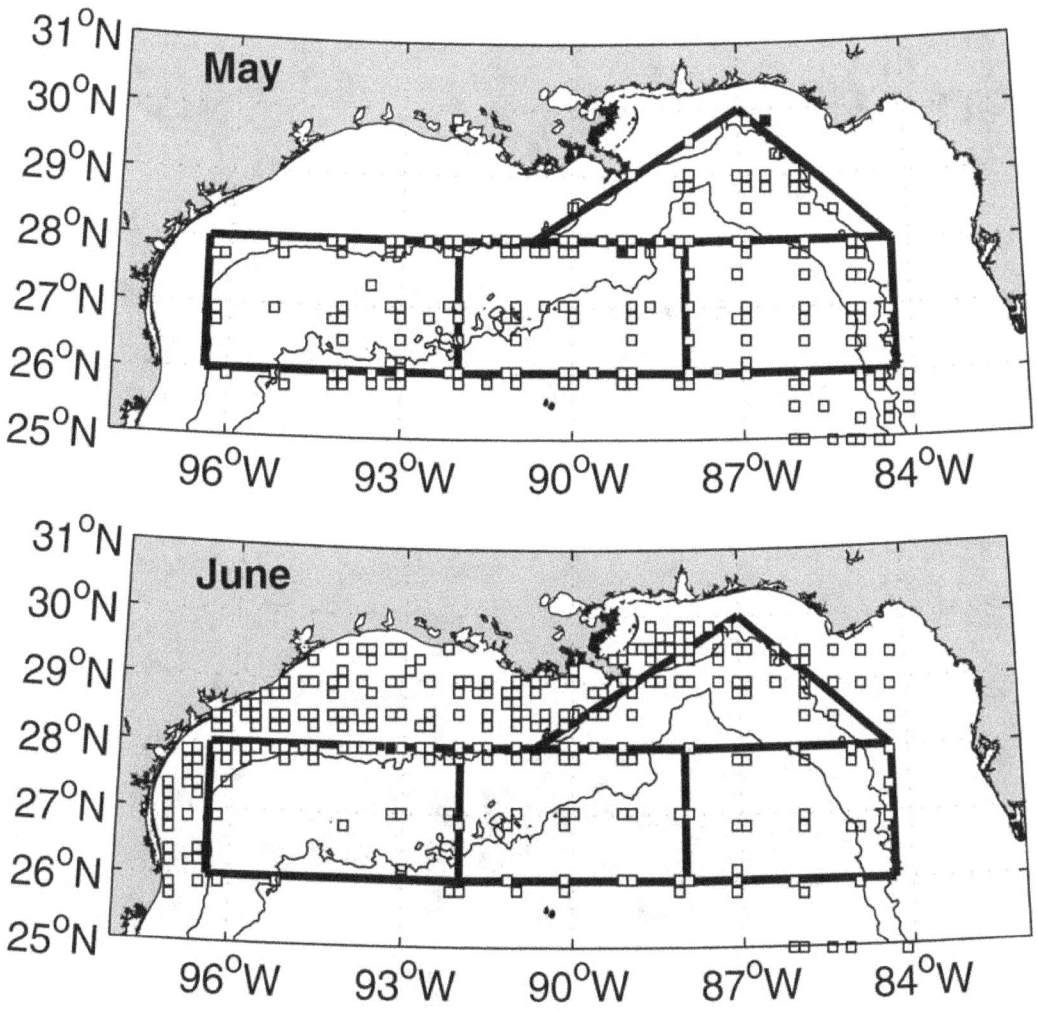

Figure 9. Presence (■) and absence (□) of yellowfin tuna larvae in the study area during May and June determined from SEAMAP ichthyoplankton data. Confirmed larvae were only present during May and June.

4.1.4 Predicted Adult Distributions

Adult yellowfin tuna are likely present throughout the majority of the study area during all months of the year (Fig. 10). During winter (January through March), the majority of confirmed landings were in the western and western halves of the central and northern zones. By April, yellowfin expand their distribution in the northern zone towards the northeast and this movement pattern continues through May and June (Fig. 10). During April there is also an apparent movement into waters deeper than 200 m in the southeastern edge of the northern zone and the northeastern edge of the eastern zone. In summer, adults are present throughout most of the three zones and during fall and early winter (September through December), the epicenter of confirmed records shifts back to the western and central zones (Fig. 10).

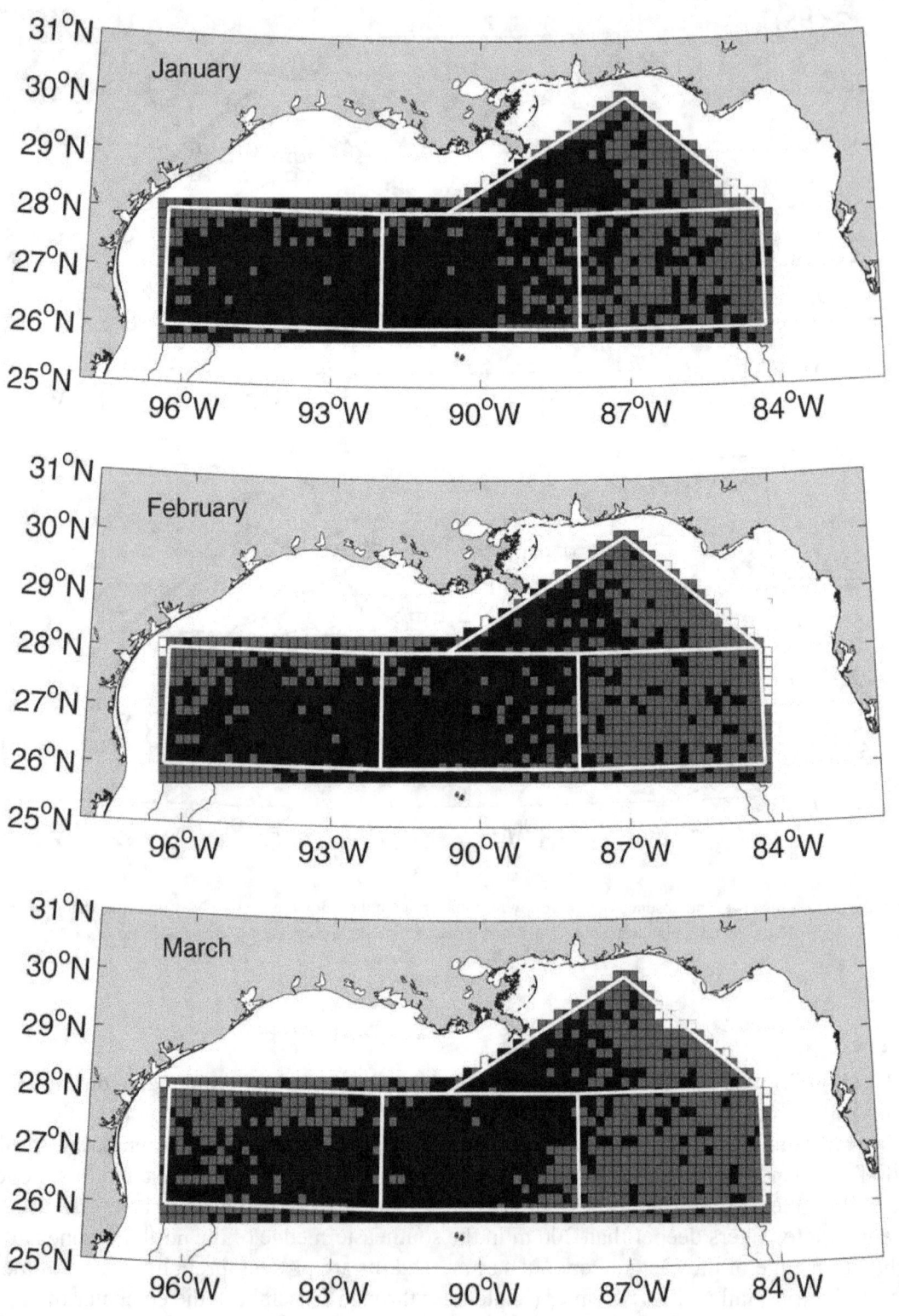

Figure 10. Predicted distributions of adult yellowfin tuna in the study area from January through December. The presence of individuals in each grid cell is coded as: confirmed (■), reasonable inference (■) or unreported (□).

32

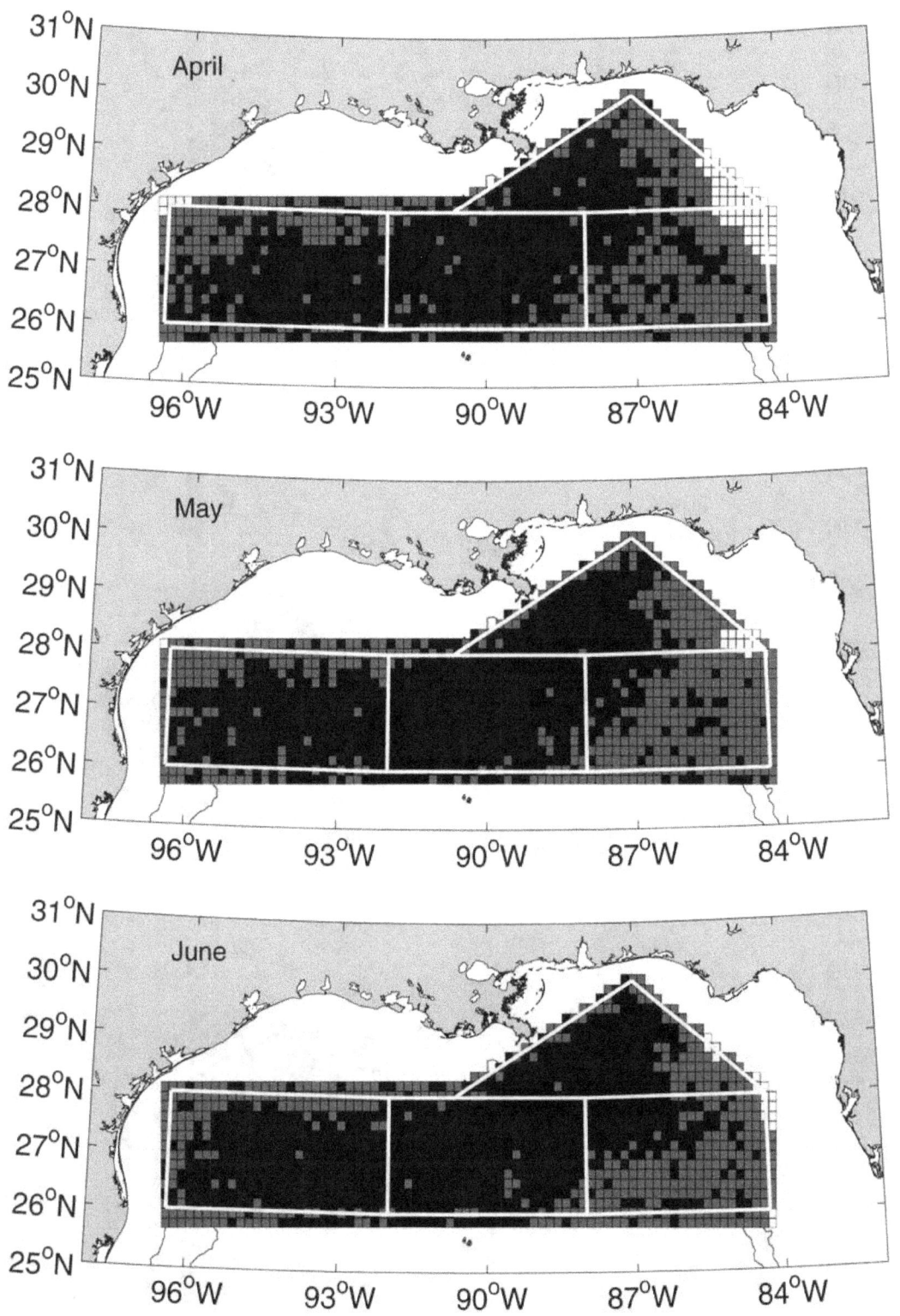

Figure 10. Predicted distributions of adult yellowfin tuna in the study area from January through December. The presence of individuals in each grid cell is coded as: confirmed (■), reasonable inference (▦) or unreported (□). (continued)

33

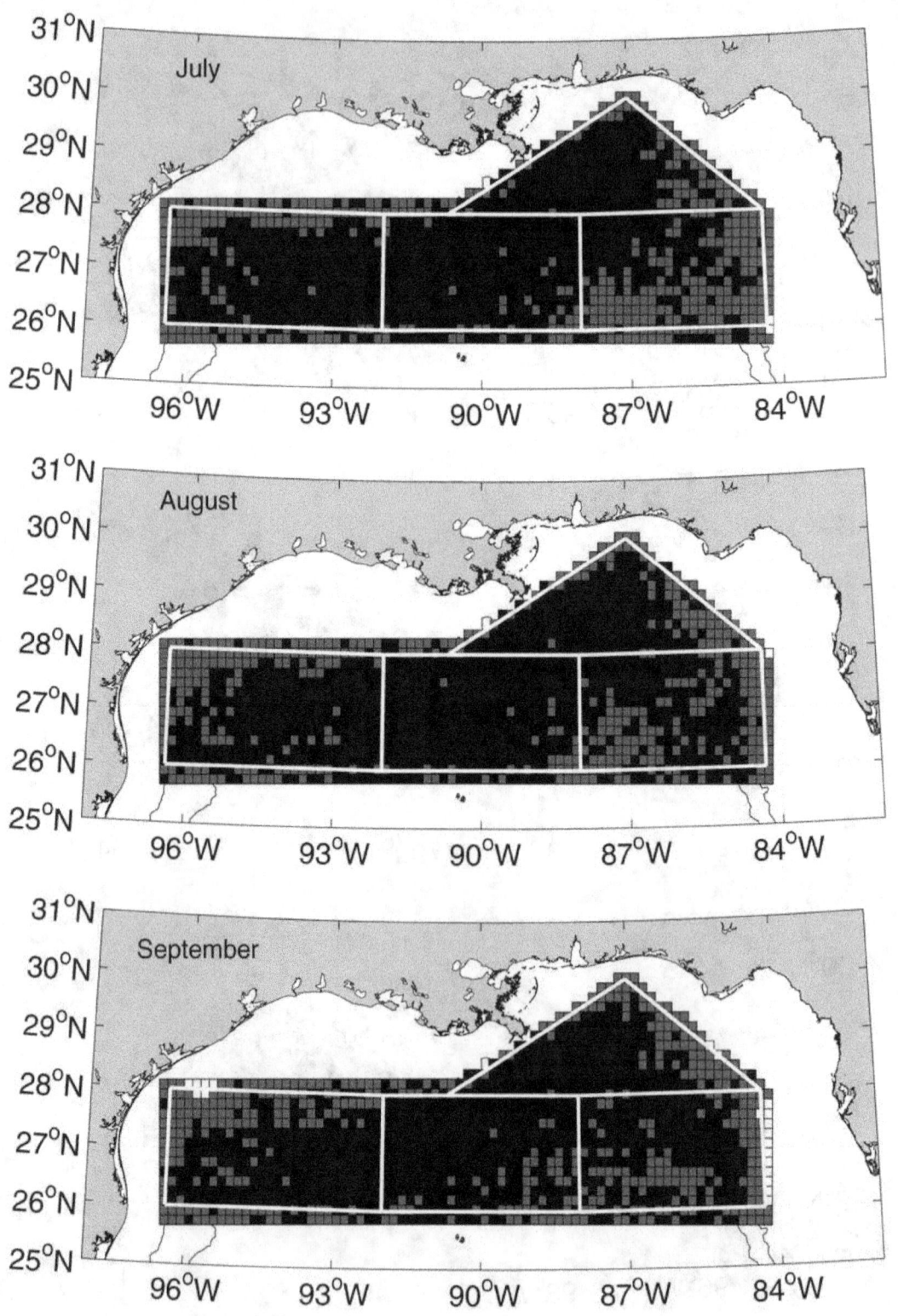

Figure 10. Predicted distributions of adult yellowfin tuna in the study area from January through December. The presence of individuals in each grid cell is coded as: confirmed (■), reasonable inference (■) or unreported (□). (continued)

34

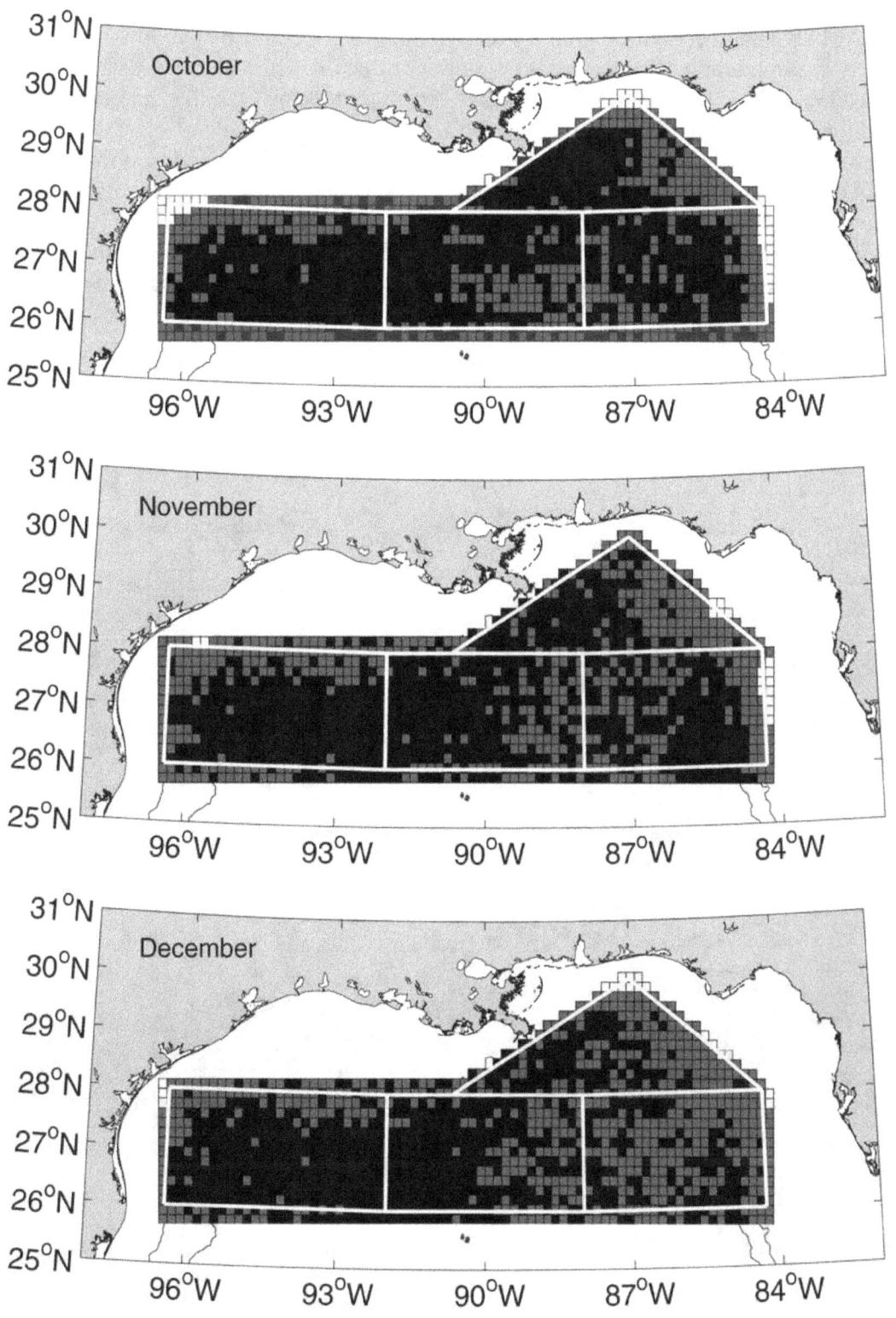

Figure 10. Predicted distributions of adult yellowfin tuna in the study area from January through December. The presence of individuals in each grid cell is coded as: confirmed (■), reasonable inference (▩) or unreported (☐). (continued)

4.1.5 Predicted Larval/Juvenile Distributions

The SEAMAP dataset contains extremely limited numbers of confirmed yellowfin larvae that were only present during May and June. Predictions of larval distributions based on this dataset are restricted to these two months (Fig. 11) and do not provide much utility for estimating larval distributions. Additional data from Grimes and Lang (1992) indicated larvae off the Mississippi River plume during September (Fig. 8). When the larval distributions predicted from the SEAMAP and Grimes and Lang (1992) data are viewed together, they suggest that most spawning occurs near the Mississippi River plume frontal region with larval and juvenile yellowfin tuna present seaward and downstream (southwest) of the plume along the 200 m (Fig. 12). In the absence of juvenile distributional data, it is likely that their distributions overlap with the larvae.

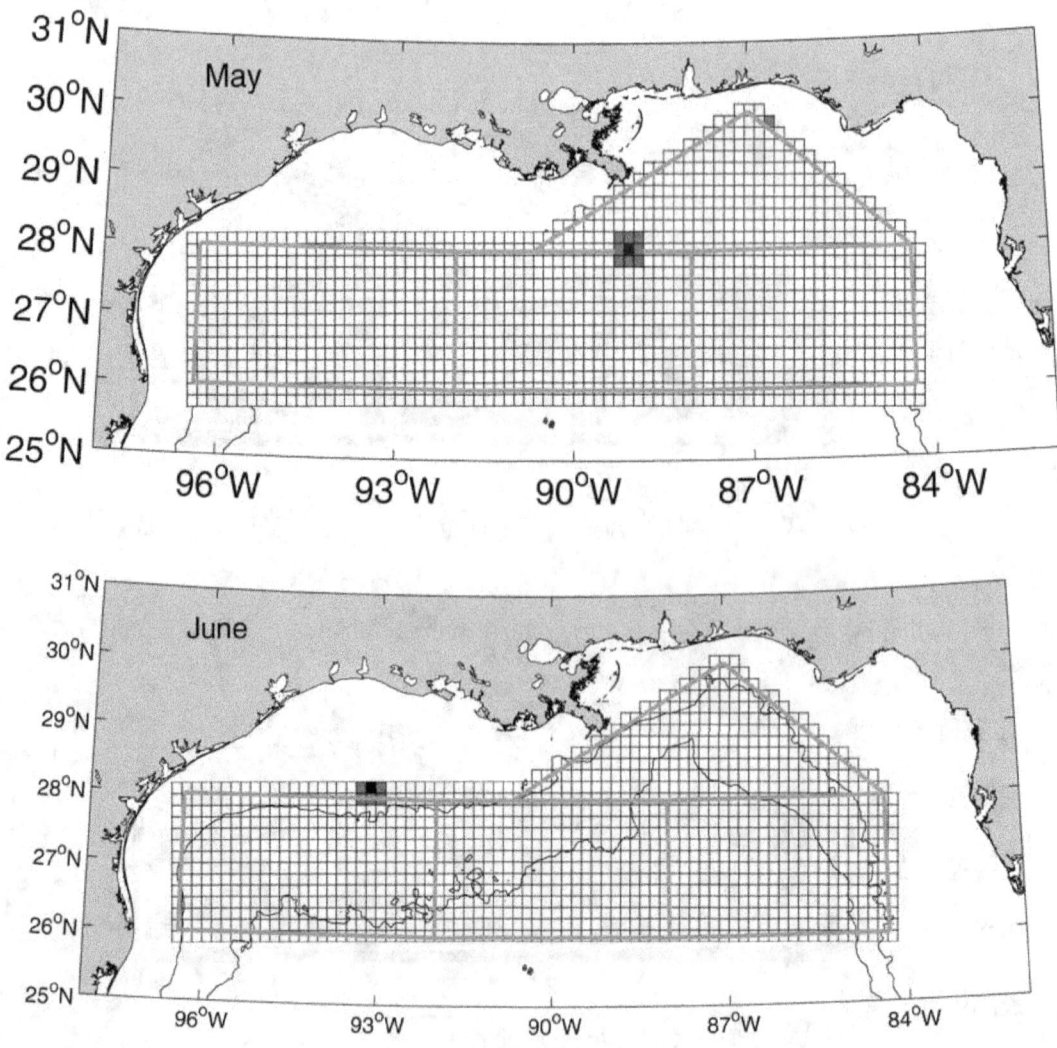

Figure 11. Predicted distributions of larval/juvenile yellowfin tuna in the study area during May, June, and September. The presence of individuals in each grid cell are indicated as (■), predicted (▨) or unreported (□).

36

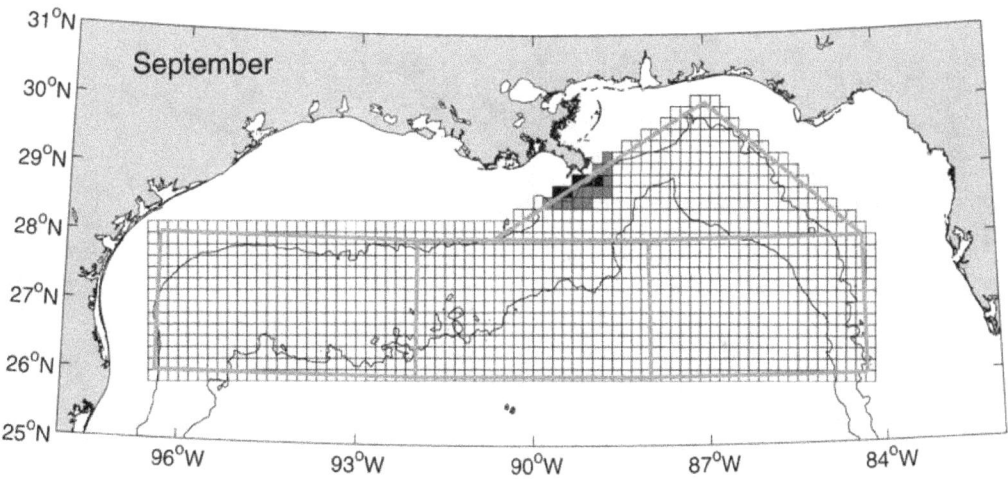

Figure 11. Predicted distributions of larval/juvenile yellowfin tuna in the study area during May, June, and September. The presence of individuals in each grid cell are indicated as (■), predicted (▨) or unreported (□). (continued)

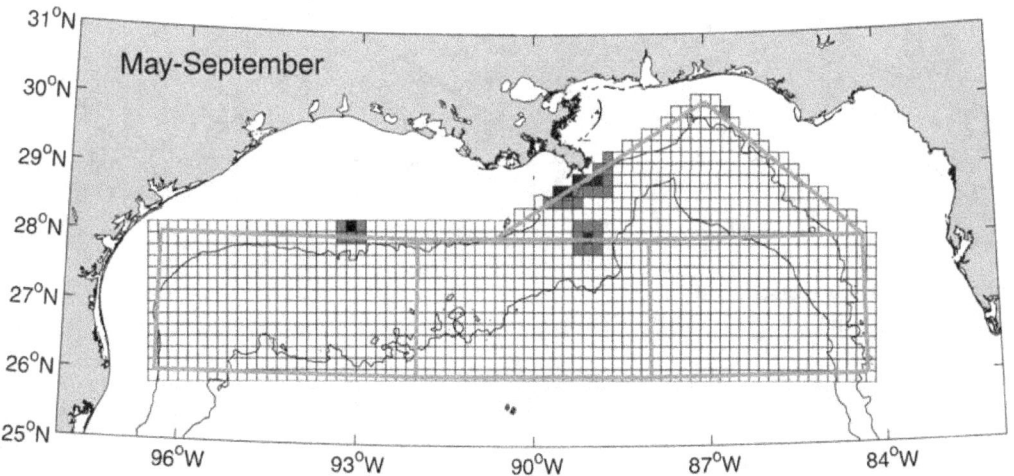

Figure 12. Predicted distributions of larval/juvenile yellowfin tuna in the study area during summer based on a composite from May, June and September. The presence of individuals in each grid cell are indicated as (■), predicted (▨) or unreported (□).

4.2 Bluefin Tuna (*Thunnus thynnus*)
4.2.1 Adult Distributions
Honma et al. (1985) indicated that adult bluefin tunas were most abundant within an area bounded by 25 to 30° and -95 to -85° from April to May. Temperature was considered an important determinant of the distributions of adult bluefin tunas (Maul et al. 1984). A seasonal pattern of distribution based on the long-line fishing effort provides an estimate of the distributional range of this species in the area of interest (Fig. 13). Adults were most abundant in the long-line fishery from January through May (Fig. 13). NOS (1985) predicts adults to be present throughout most of the waters seaward of the 200 m isobath during spring (Fig. 14) and well seaward of the 2000 m isobath from winter to spring (Fig. 14).

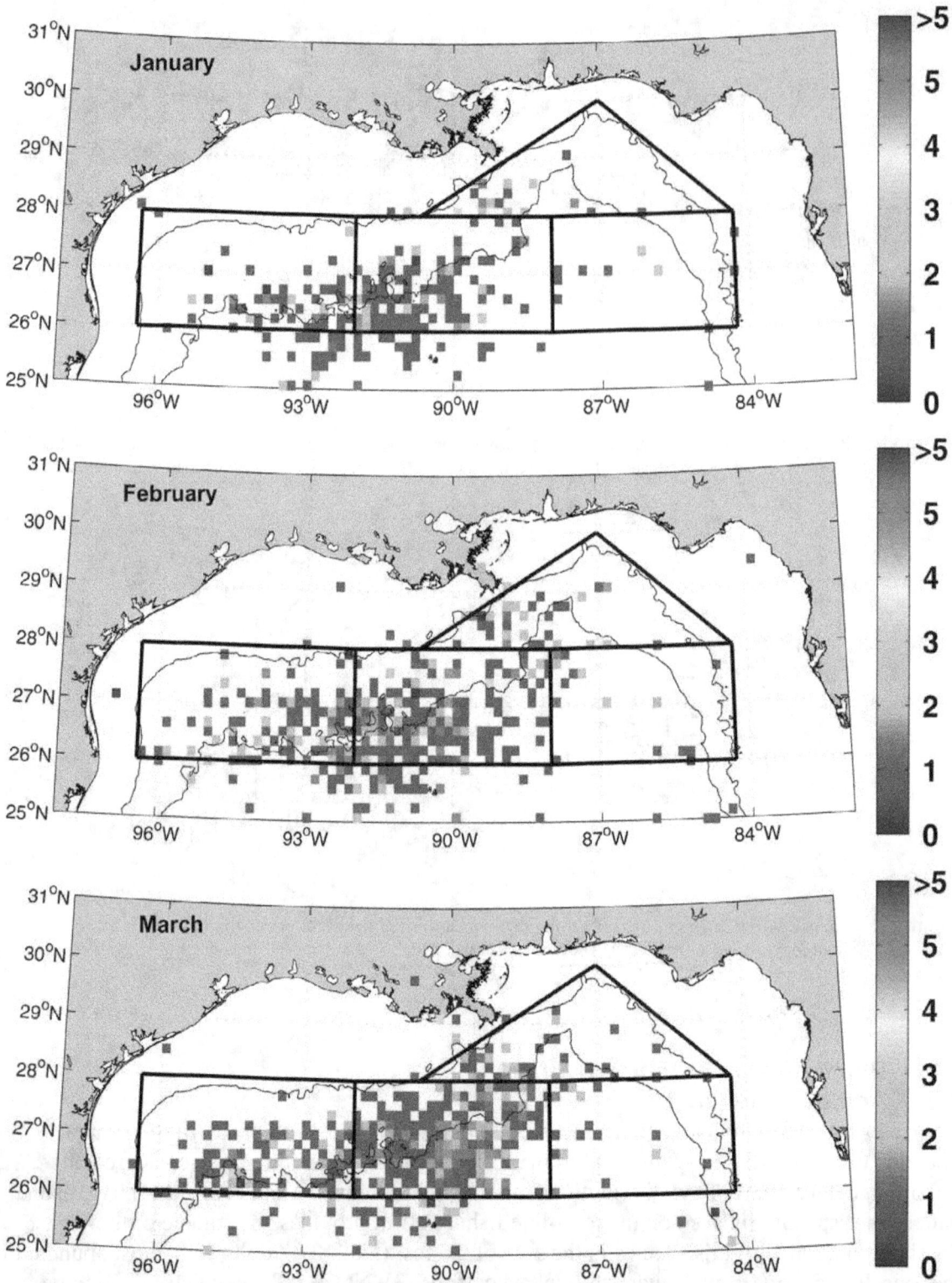

Figure 13. Catch per unit effort of adult bluefin tuna from the commercial long-line fishery. Each square represents the mean catch-per-unit-effort (tuna per set) taken within a 10' x10' region from January through December over the period 1986-1999. Note that the colorbar was arbitrarily limited to a maximum CPUE of five. The maximum CPUEs were: January = 34; February = 28; March = 15.5; April = 18; May = 62;. June = 36.5; July = 42; August = 51; September = 21; October = 40; November = 6; December = 38.

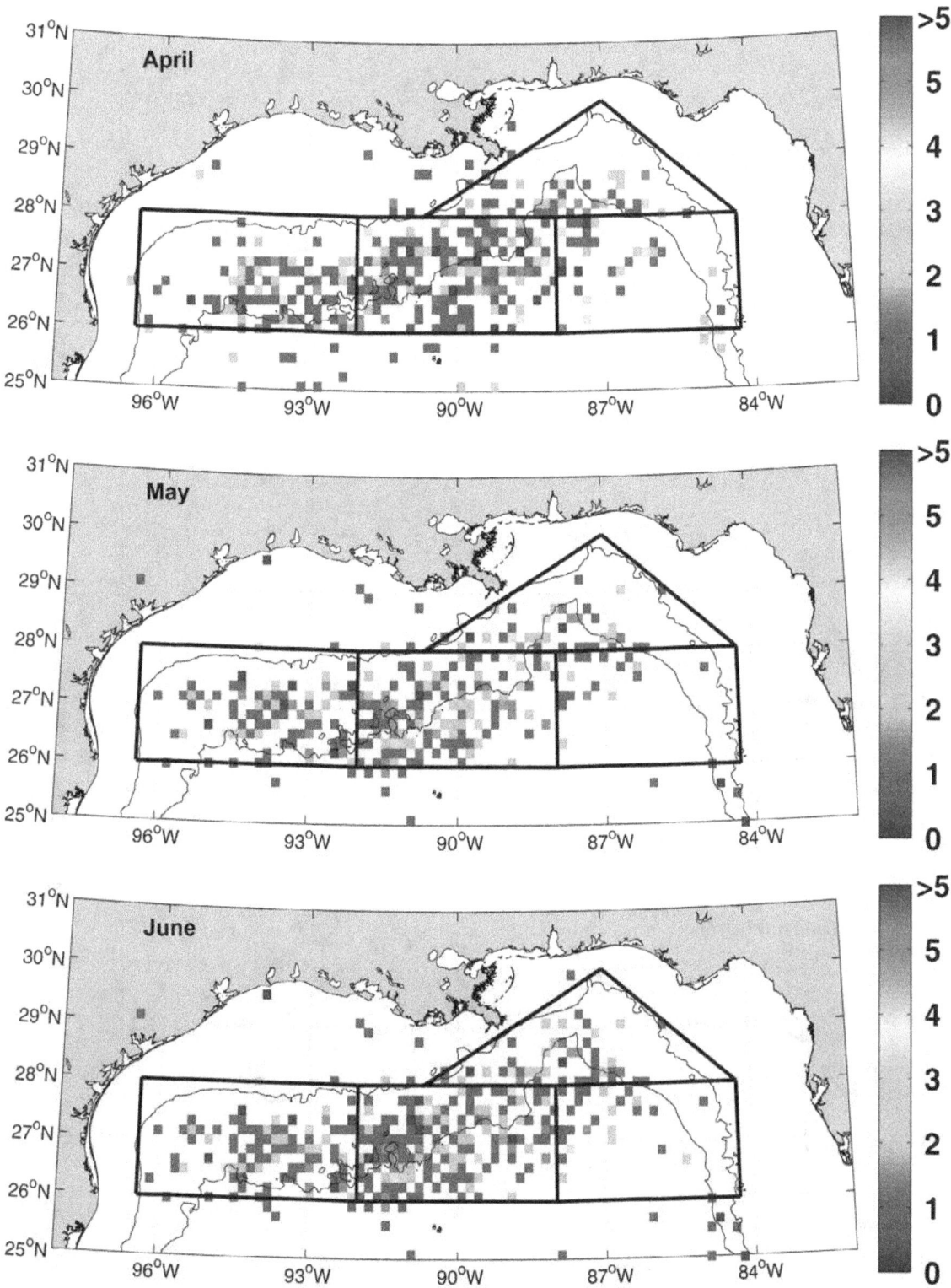

Figure 13. Catch per unit effort of adult bluefin tuna from the commercial long-line fishery. Each square represents the mean catch-per-unit-effort (tuna per set) taken within a 10' x10' region from January through December over the period 1986-1999. Note that the colorbar was arbitrarily limited to a maximum CPUE of five. The maximum CPUEs were: January = 34; February = 28; March = 15.5; April = 18; May = 62;. June = 36.5; July = 42; August = 51; September = 21; October = 40; November = 6; December = 38. (continued)

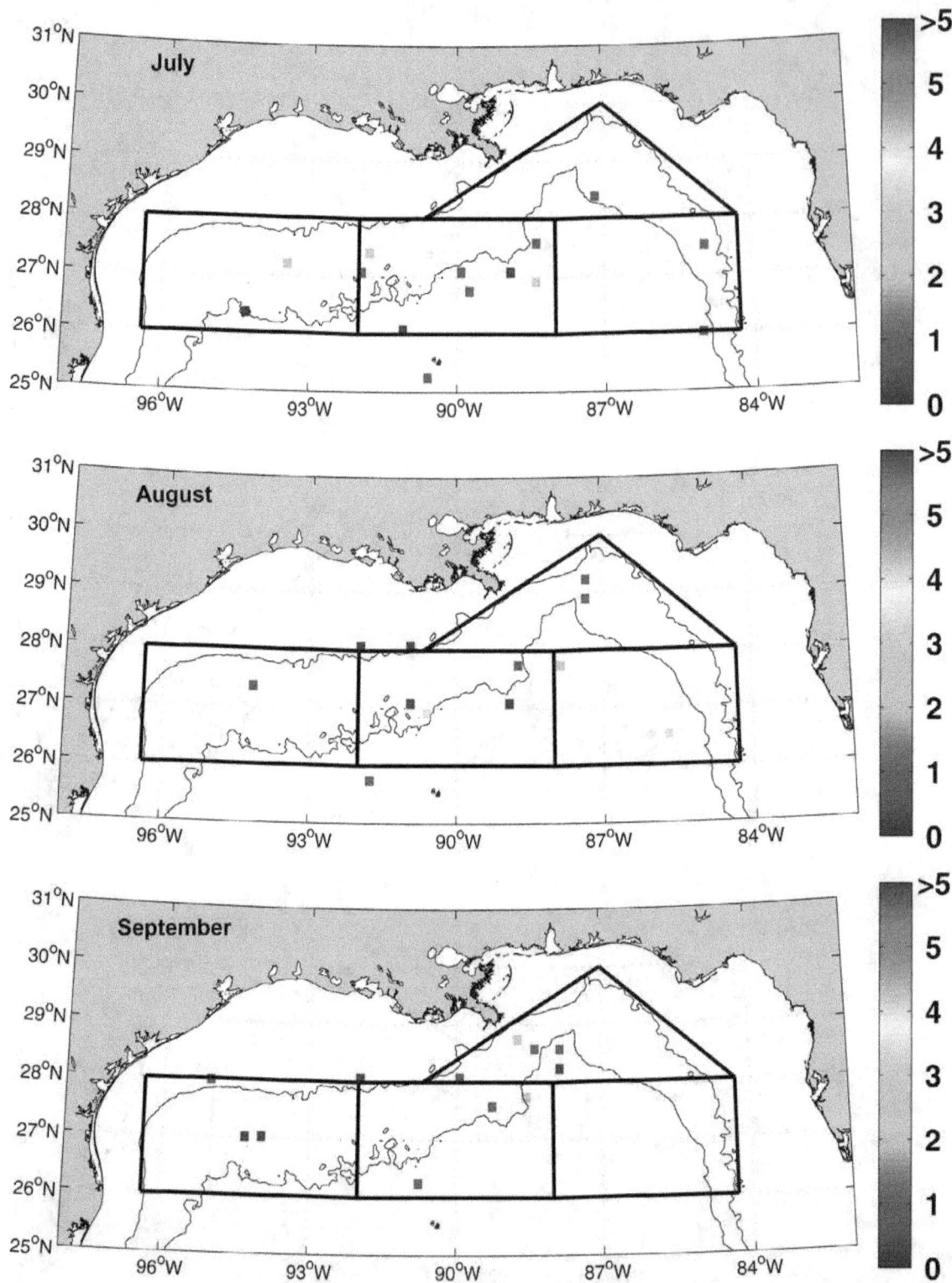

Figure 13. Catch per unit effort of adult bluefin tuna from the commercial long-line fishery. Each square represents the mean catch-per-unit-effort (tuna per set) taken within a 10' x10' region from January through December over the period 1986-1999. Note that the colorbar was arbitrarily limited to a maximum CPUE of five. The maximum CPUEs were: January = 34; February = 28; March = 15.5; April = 18; May = 62;. June = 36.5; July = 42; August = 51; September = 21; October = 40; November = 6; December = 38. (continued)

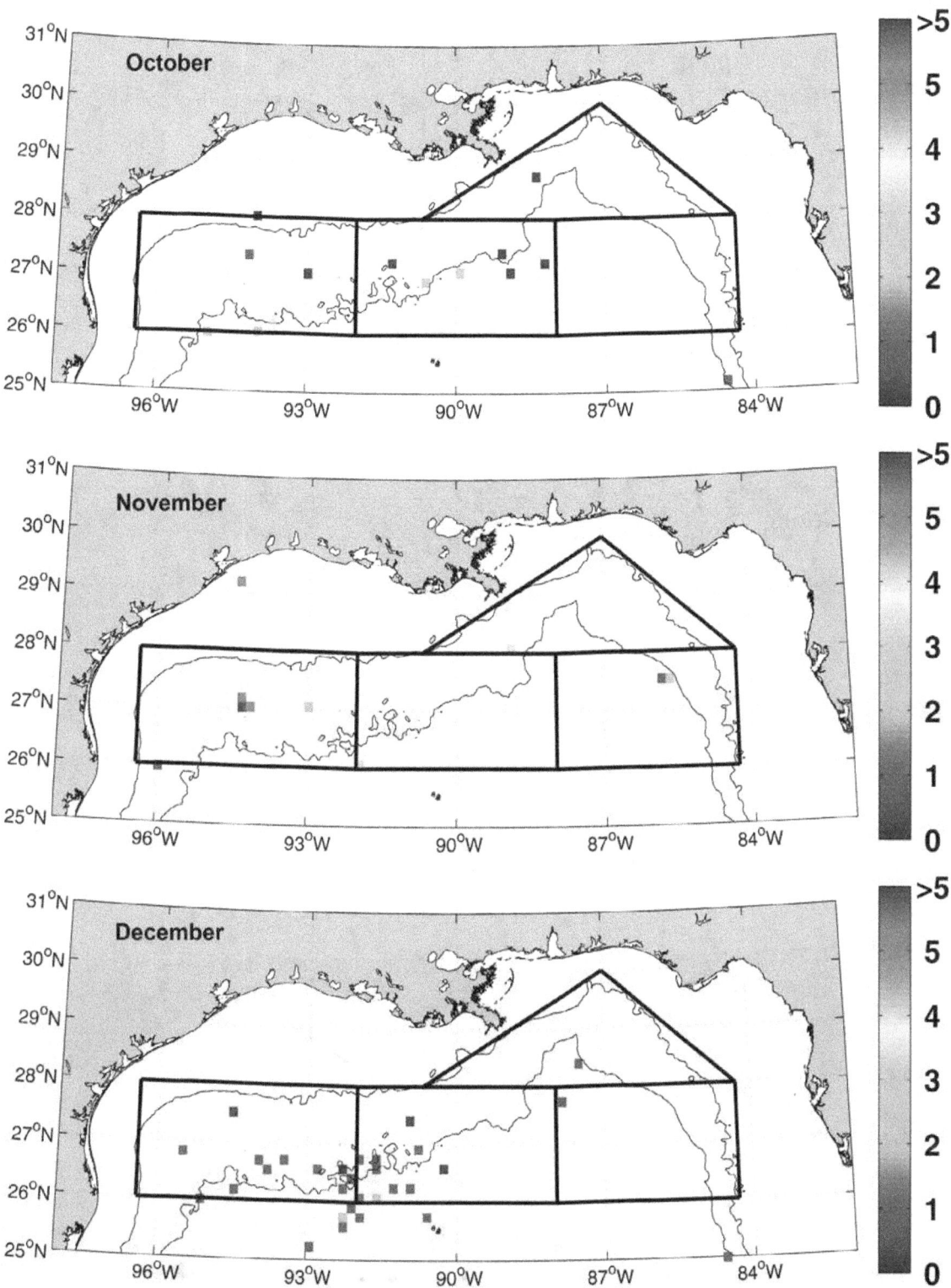

Figure 13. Catch per unit effort of adult bluefin tuna from the commercial long-line fishery. Each square represents the mean catch-per-unit-effort (tuna per set) taken within a 10' x10' region from January through December over the period 1986-1999. Note that the colorbar was arbitrarily limited to a maximum CPUE of five. The maximum CPUEs were: January = 34; February = 28; March = 15.5; April = 18; May = 62;. June = 36.5; July = 42; August = 51; September = 21; October = 40; November = 6; December = 38. (continued)

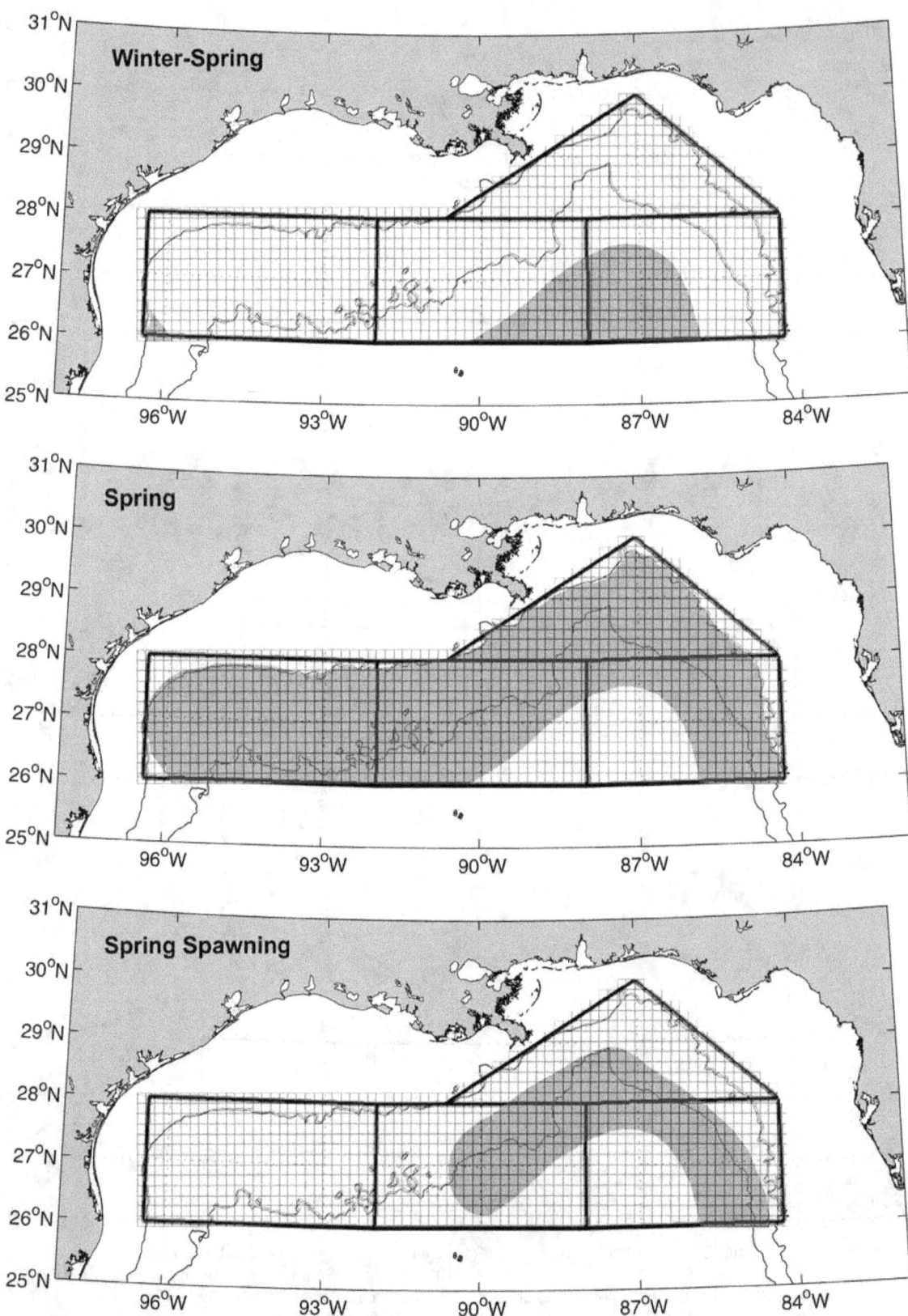

Figure 14. Distributions of adult bluefin tuna predicted by NOS (1985).

4.2.2 Reproduction

Western Atlantic bluefin tunas spawn in the Gulf of Mexico during April, May and June (Richards, 1975). Females spawn after reaching eight years of age (Southeast Fisheries Science Center, 1992). Spawning in the Gulf of Mexico occurs during April, May and June (Baglin Jr., 1982) based on seasonal variation in the gonadosomatic index of female tunas. Ditty et al. (1988) reported the presence of larvae in the northern Gulf of Mexico during April, May and June thus supporting this periodicity. Cramer and Scott (1997) considered spawning to occur during May in the Gulf of Mexico. Richards et al. (1989) stated that the bluefin spawning season extends from April 15th to June 15th. NOS (1985) indicate that spawning occurs in a region broadly centered on the 2000 m isobath between approximately 84 and 90 °W (Fig. 14).

4.2.3 Larval/Juvenile Distributions

Bluefin tuna larvae are rare in plankton tows and standard double oblique bongo net sampling tends to undersample surface waters where larvae are concentrated (McGowan and Richards, 1989). Larvae are highly motile and grow rapidly, which further contributes to avoidance of nets and underestimates of abundance (McGowan and Richards, 1989). Ditty et al. (1988) reported the presence of larvae in the northern Gulf of Mexico during April, May and June. Scott et al. (1993) developed indices of the abundances of larval bluefin tunas using Gulf of Mexico ichthyoplankton survey data collected from 1977-88. They standardized catches from bongo net casts and neuston net samples, adjusted for differences in gear efficiency, and developed an annual index of larval abundance for each sampling station.

In the Gulf of Mexico, NMFS surveys indicated that bluefin tuna larvae were present where sea surface temperatures ranged from 22.0-28.1 °C (McGowan and Richards, 1989). They concentrate in areas where currents or eddies encounter the shelf between the 100 and 1000 m isobaths (Sherman et al., 1983) and along the cold edge of the Loop Current in the eastern Gulf (McGowan and Richards, 1989; Richards et al. 1989). Surveys during 1982 and 1983 indicated that larvae were generally offshore of the 200 m isobath (McGowan and Richards, 1986; Fig. 15). SEAMAP surveys detected larvae during April, May and June (Fig. 16) in regions generally in waters above the 200 and 2000 m isobaths.

Larval densities are highly variable. Mather et al. (1995) found larvae present at densities from 0–2000 larvae 100 m^2 at stations in the Gulf of Mexico during April and May. All stations with larval *T. thynnus* present were between 23-28.5°N and 94.5-84.5°W and were seaward of the 200 m isobath. Larvae were more abundant in Gulf waters during May and June than during April (Mather et al., 1995). While much less abundant during July, bluefin larvae were collected during July in the northern Gulf near 29°N, 87°W (Mather et al. 1995). Little is reported on the distributions of juvenile bluefin tunas. According to Mather et al. (1995), the range of small juvenile bluefin tuna encompasses that of the larvae in the Gulf and may extend further north.

Little can be inferred about juvenile bluefin tuna distributions from the SEAMAP data. The maximum length of specimens collected during this period was only 8.2 mm. Tuna grow rapidly and become competent swimmers, which likely contributes to effective avoidance of gear designed to collect ichthyoplankton. The distributions of early juvenile bluefin tunas probably overlap that of the larvae.

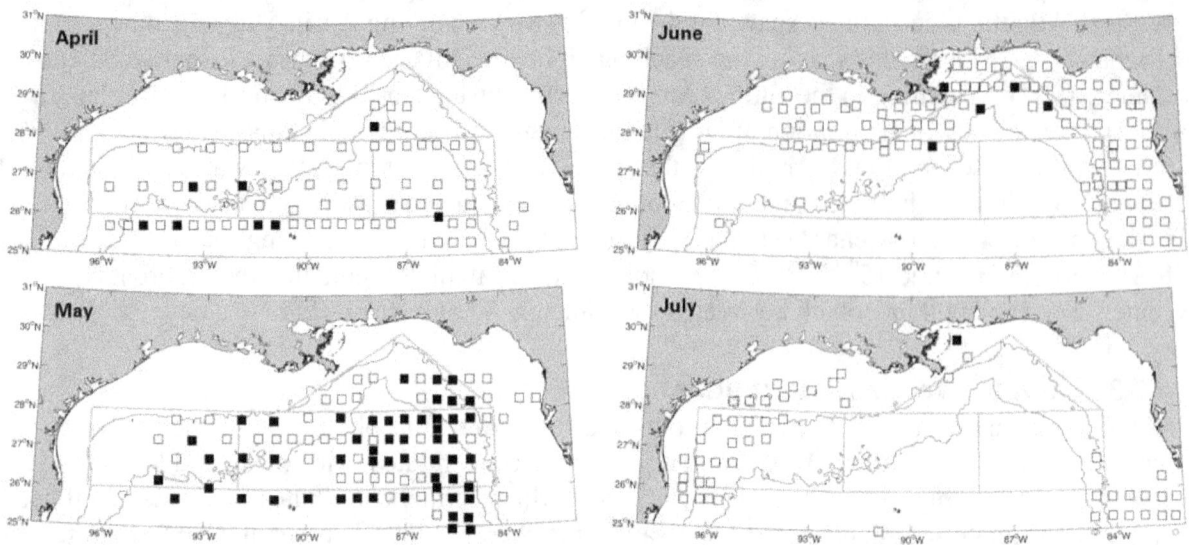

Figure 15. Presence (■) and absence (□) of larval bluefin tuna in bongo net and neuston net samples collected in the Gulf of Mexico during April, May, June and July. Data digitized from McGowan and Richards (1986) Figs. 1–16 and Richards et al. (1993) Appendix Tables 4-9.

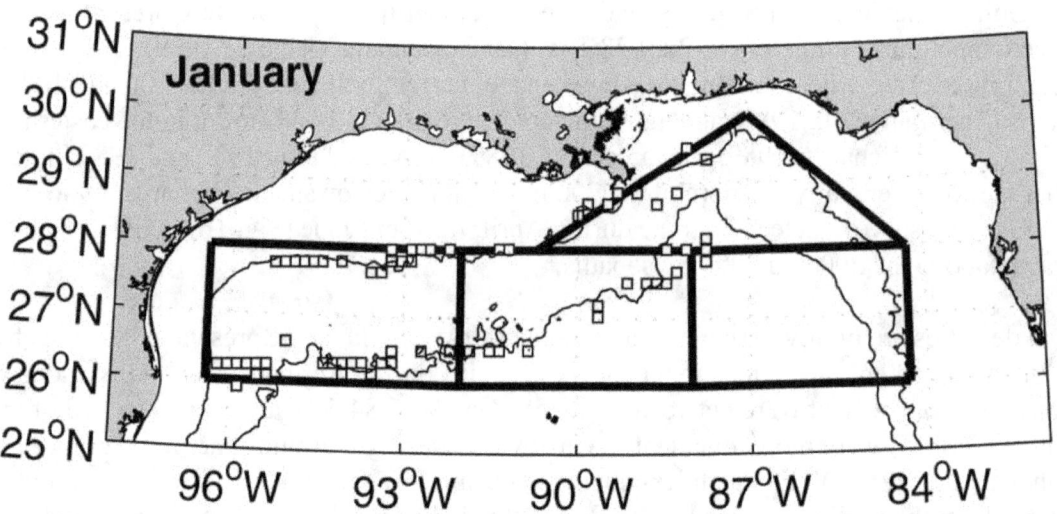

Figure 16. Presence (■) and absence (□) of bluefin tuna larvae in the study area from January through December estimated from SEAMAP ichthyoplankton data.

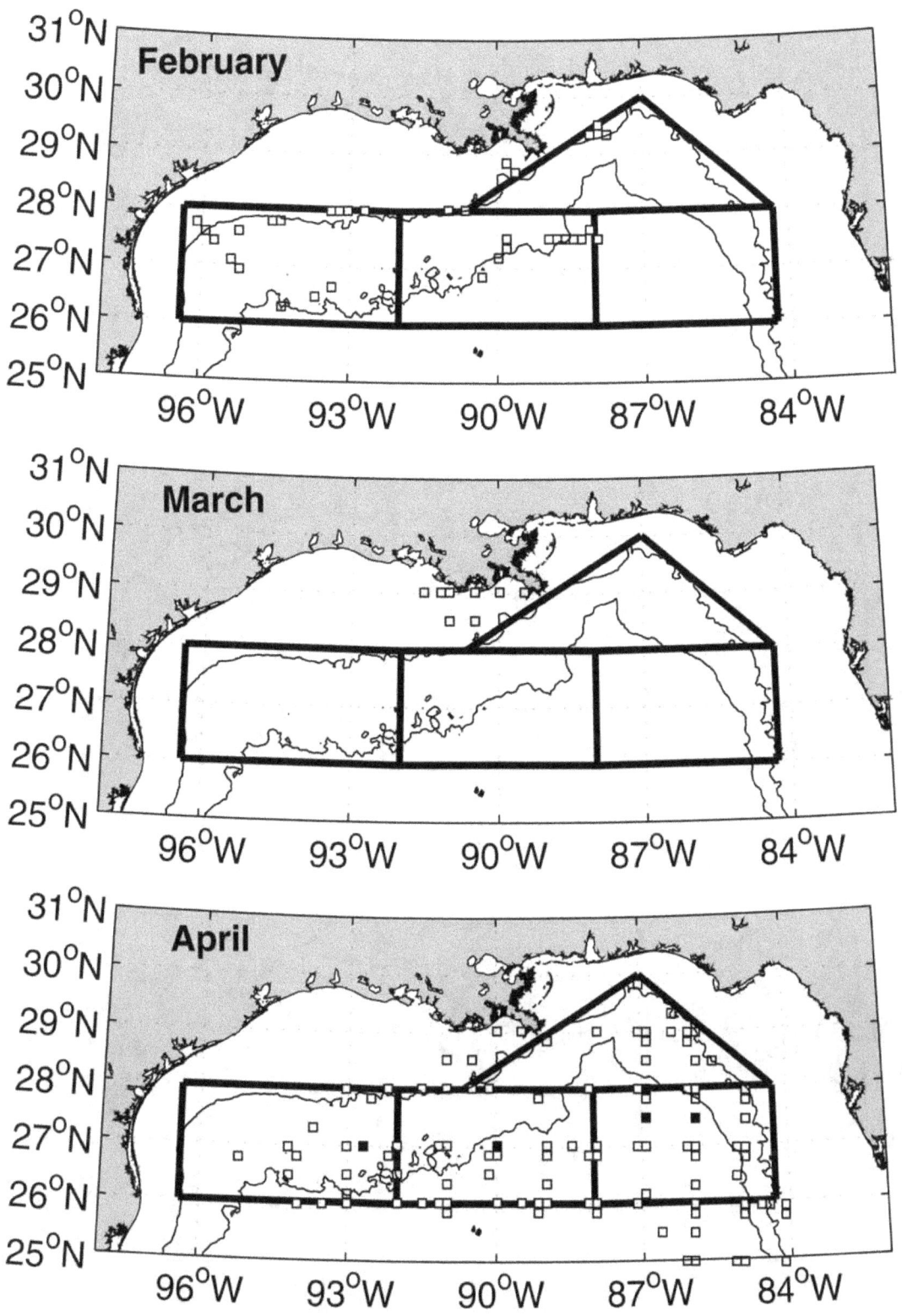

Figure 16. Presence (■) and absence (□) of bluefin tuna larvae in the study area from January through December estimated from SEAMAP ichthyoplankton data. (continued)

45

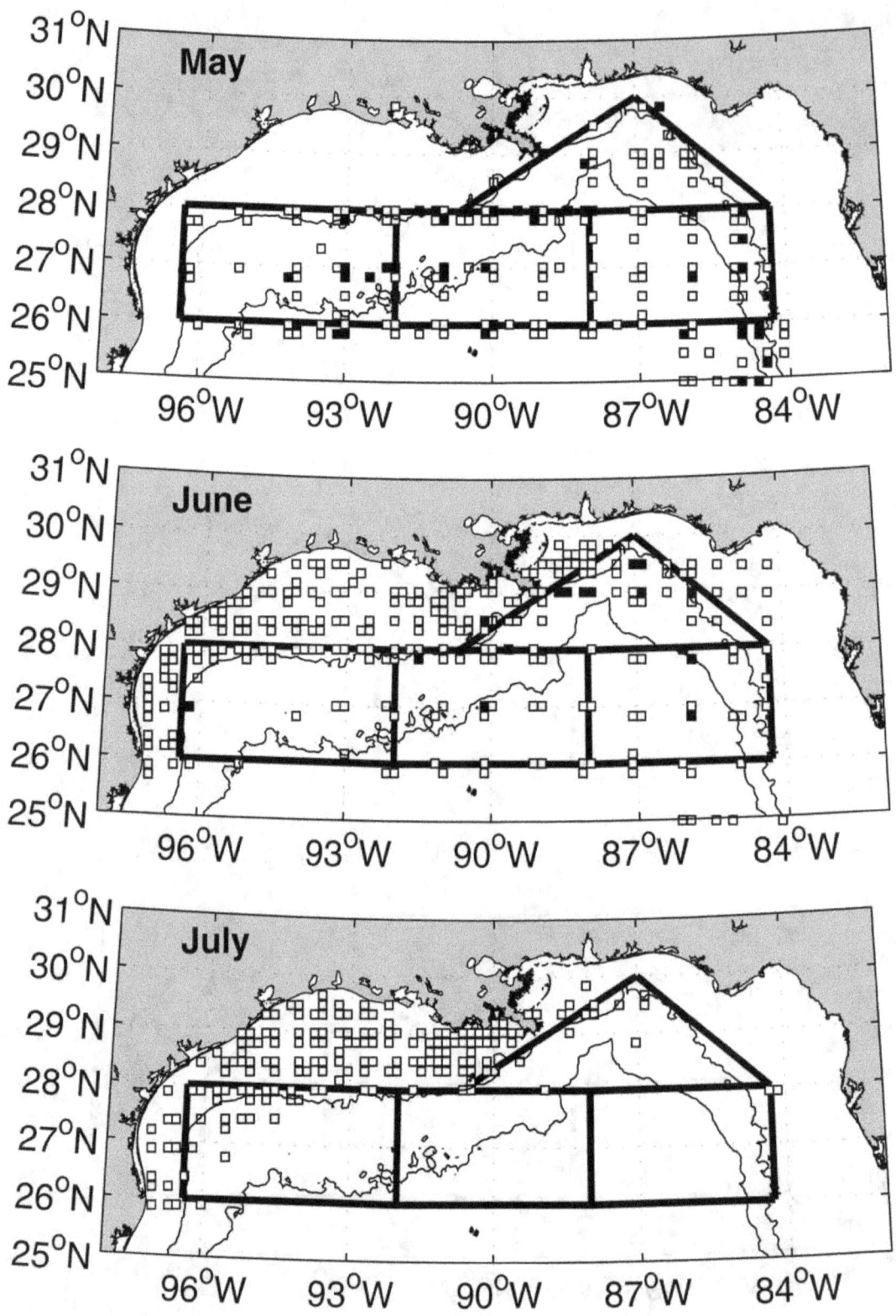

Figure 16. Presence (■) and absence (□) of bluefin tuna larvae in the study area from January through December estimated from SEAMAP ichthyoplankton data. (continued)

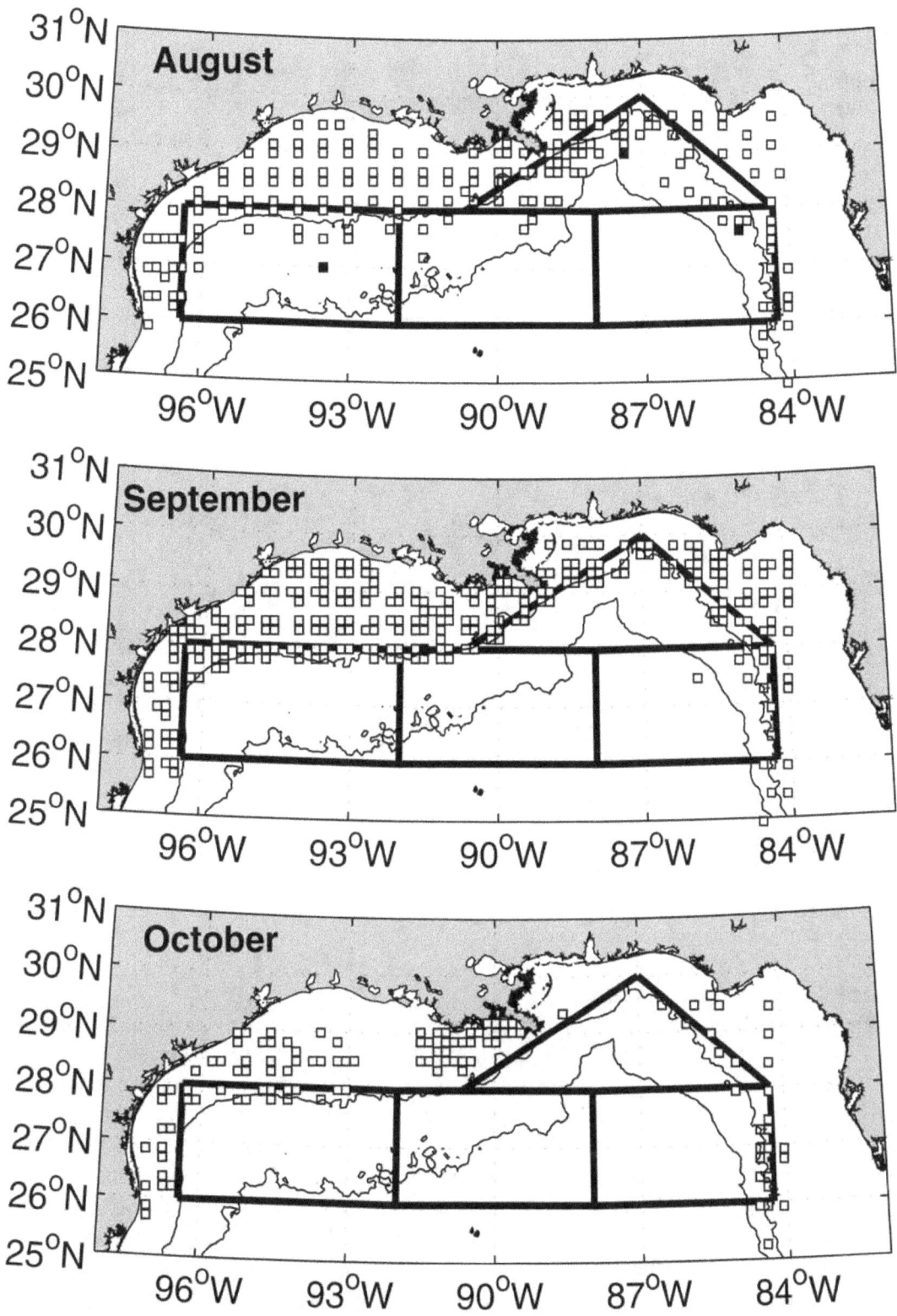

Figure 16. Presence (■) and absence (□) of bluefin tuna larvae in the study area from January through December estimated from SEAMAP ichthyoplankton data. (continued)

47

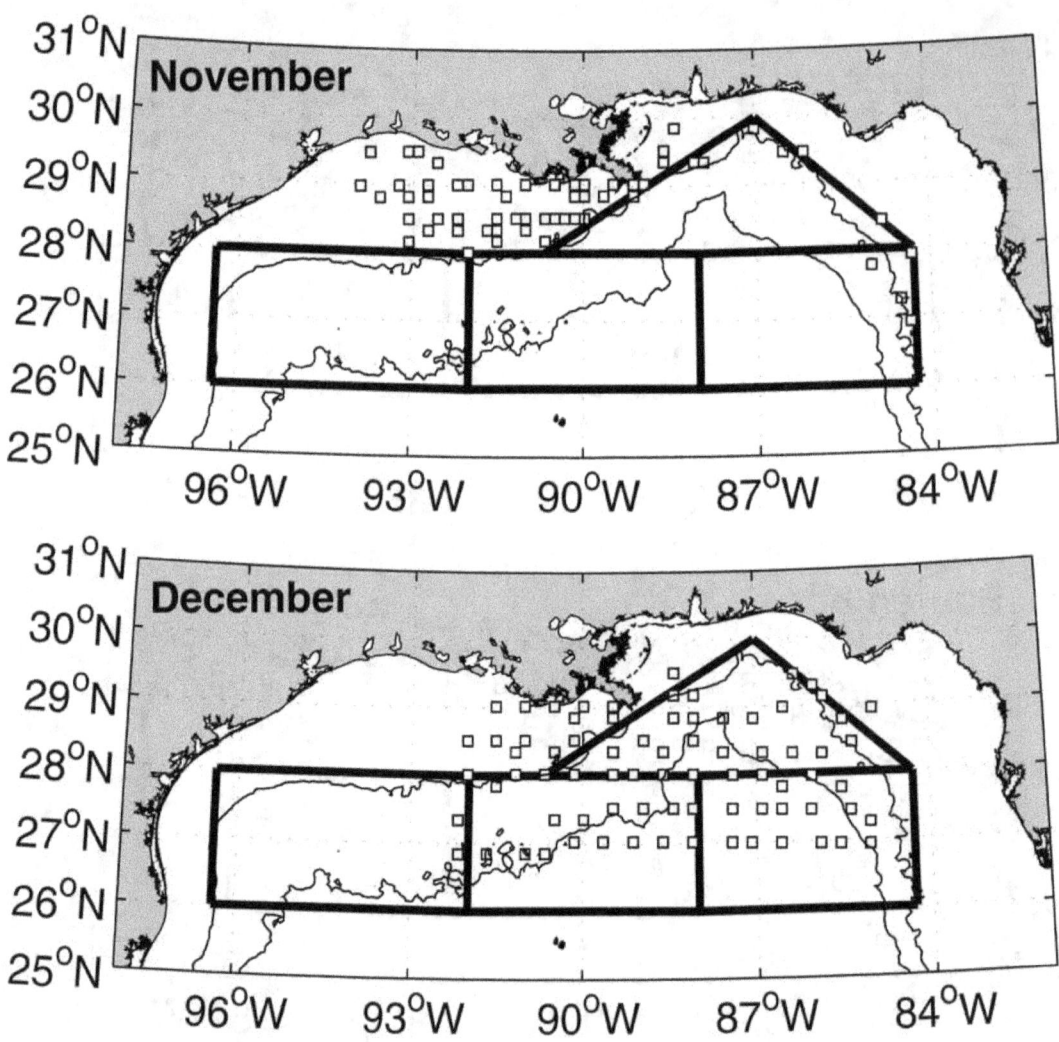

Figure 16. Presence (■) and absence (□) of bluefin tuna larvae in the study area from January through December estimated from SEAMAP ichthyoplankton data. (continued)

4.2.4 Predicted Adult Distributions

Adult bluefin tuna are likely present in the Gulf of Mexico throughout the year (Fig. 17). From January, when their distributional epicenter appears to be located in the southwestern portion of the central region of the study area, through March, when the majority of confirmed records occupies most of the central zone, bluefin expand their region of occurrence throughout most of the waters of the study area deeper than 200 m (Fig. 17). During April and May, the zone based on confirmed records expands into the western and northern zones, while becoming more diffuse. At the same time adult bluefin become less abundant in the southeastern section of the eastern zone. By June, the number of cells containing confirmed records of adult bluefin has retracted from the periphery of the eastern, western and eastern half of the northern zones (Fig. 17). From July through November, the distribution of adult bluefin diminishes until they scatter into isolated pockets (Fig. 17). During December there appears to be a reconstitution of the adult population in the southeastern region of the western zone and the southwestern region of the central zone indicating the onset of a resurgence of adult bluefin in the Gulf.

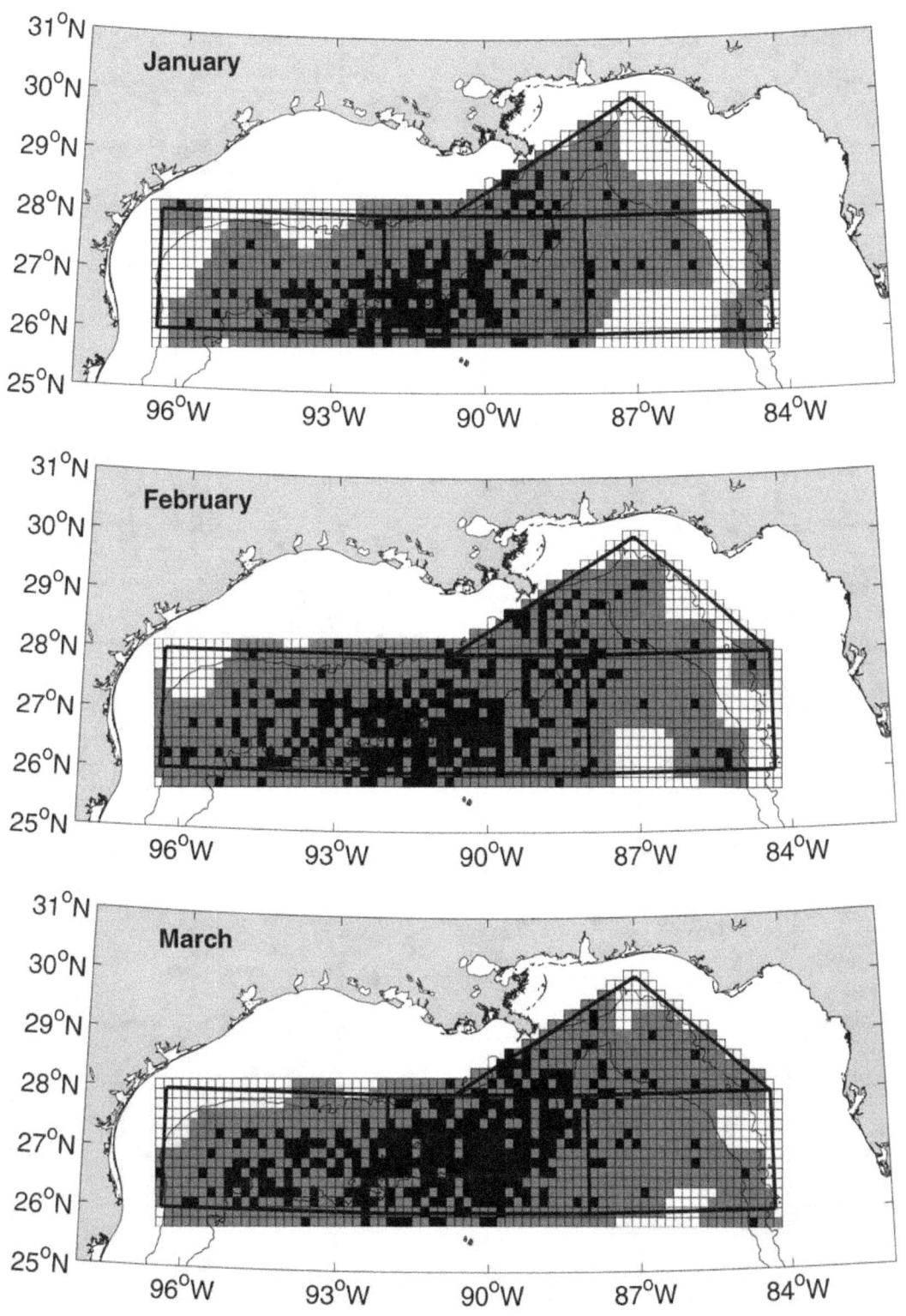

Figure 17. Predicted distributions of adult bluefin tuna in the study area from January through December. The presence of individuals in each grid cell is coded as: confirmed (■), reasonable inference (▨), or unreported (□).

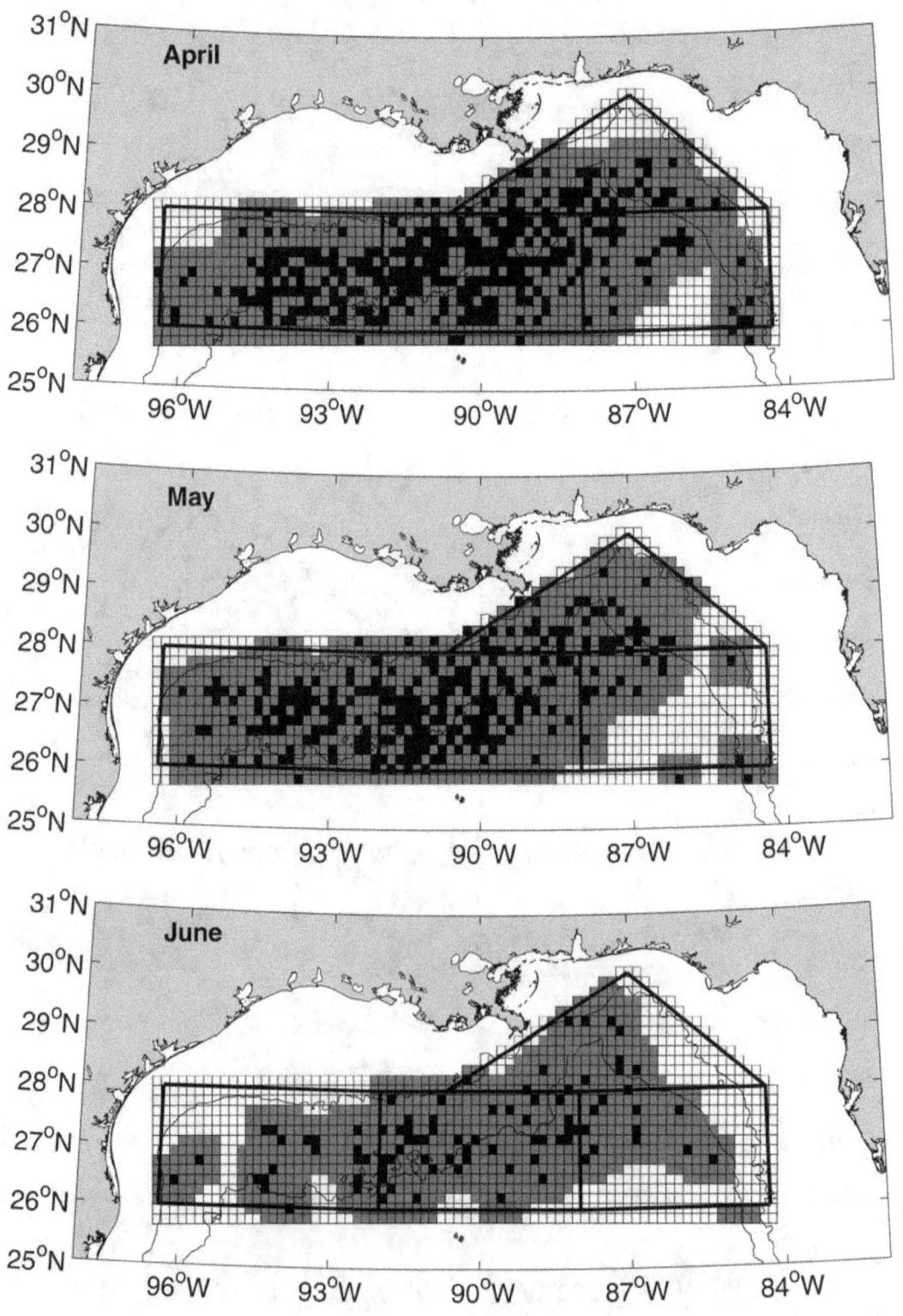

Figure 17. Predicted distributions of adult bluefin tuna in the study area from January through December. The presence of individuals in each grid cell is coded as: confirmed (■), reasonable inference (▨), or unreported (□). (continued)

50

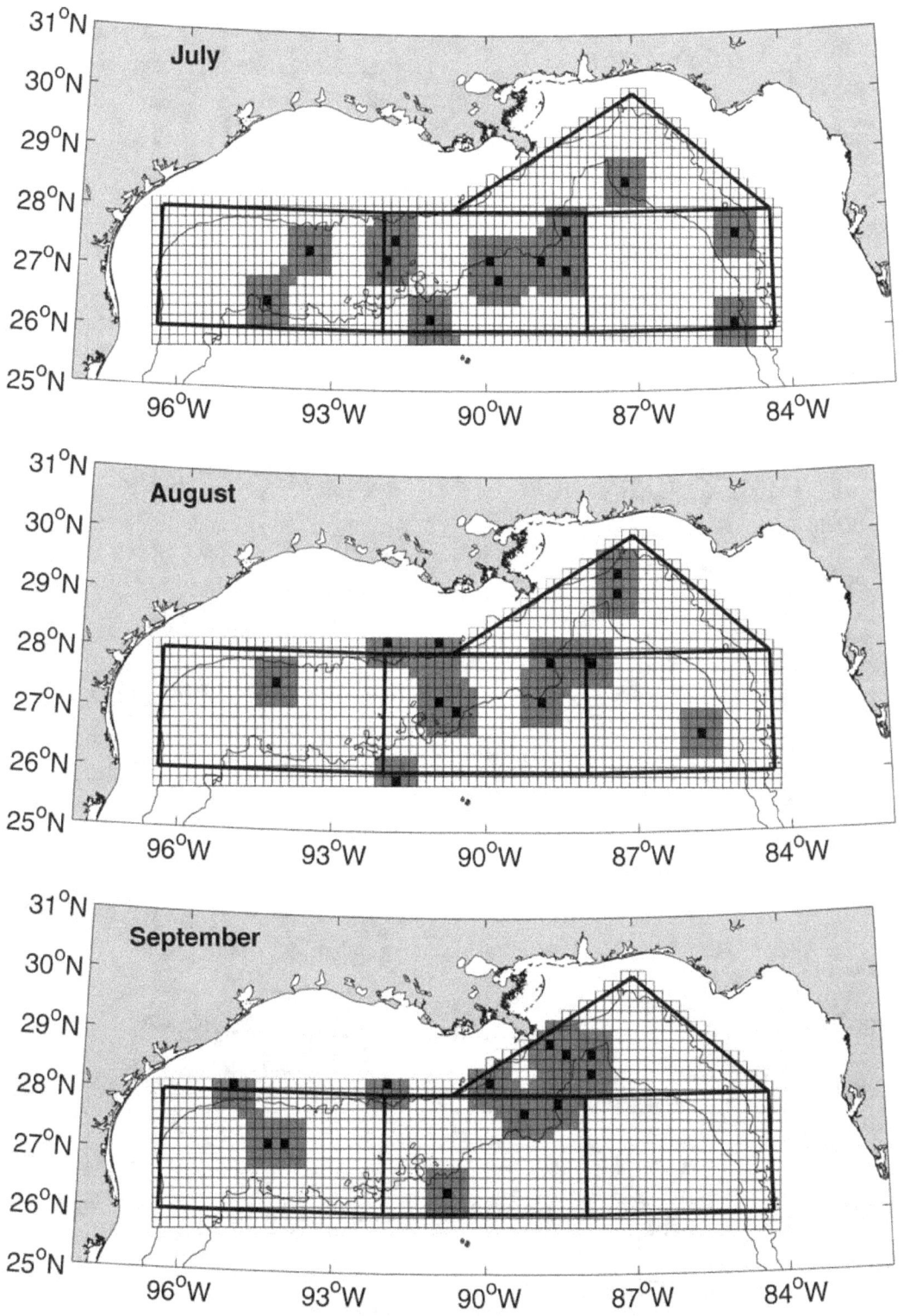

Figure 17. Predicted distributions of adult bluefin tuna in the study area from January through December. The presence of individuals in each grid cell is coded as: confirmed (■), reasonable inference (▨), or unreported (□). (continued)

51

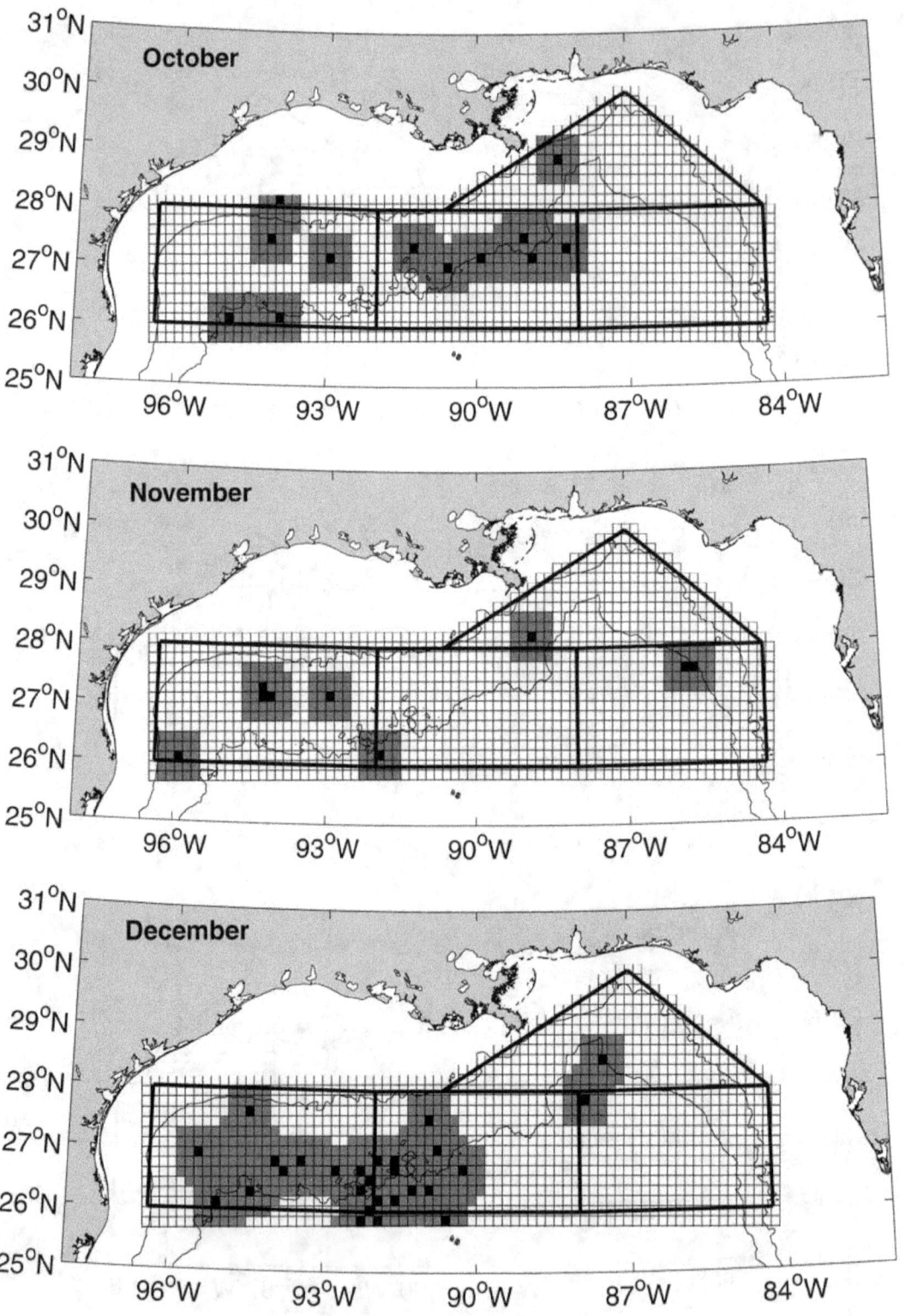

Figure 17. Predicted distributions of adult bluefin tuna in the study area from January through December. The presence of individuals in each grid cell is coded as: confirmed (■), reasonable inference (▨), or unreported (□).(continued)

4.2.5 Predicted Larval/Juvenile Distributions

Larval tuna were present in ichthyoplankton samples during April, May, June, and July, which corresponds to the reported spawning period for this species. Peak larval abundance appears to occur in May (Fig. 18). In April, larvae are present in all four study zones close to the 2000 m isobath. By May their abundance spread throughout most of the study area with the exception of the western half of the western zone, and in June their distribution was largely confined to the waters over the continental slope in the northern zone.

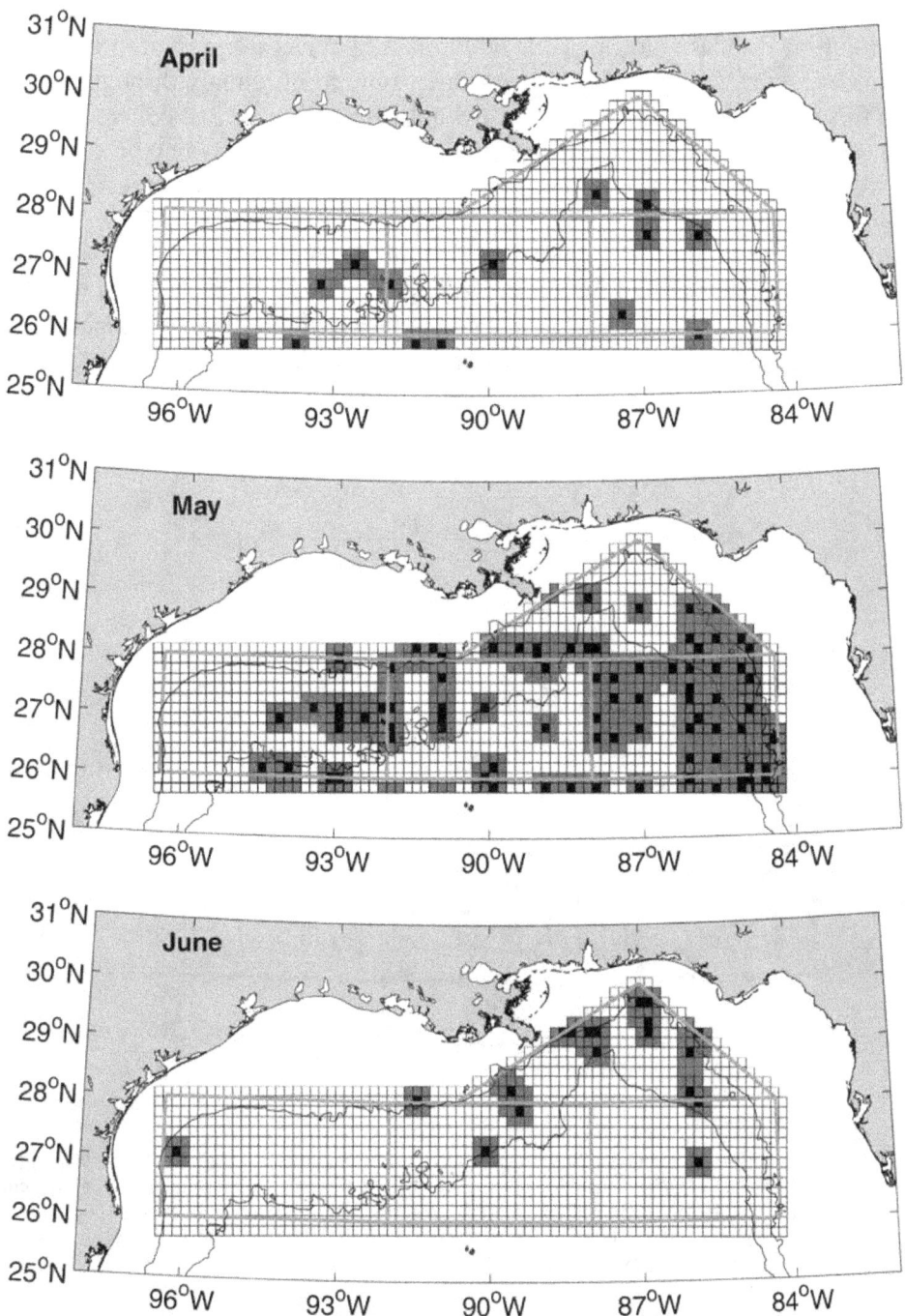

Figure 18. Predicted distributions of larval/juvenile bluefin tuna in the study area during April, May and June. The presence of individuals in each grid cell is indicated as (■), reasonable inference (▨) or unreported (□).

53

4.3   Blackfin Tuna (*Thunnus atlanticus*)
4.3.1 Adult Distributions

Blackfin tuna are a warm-water species generally found above the 20 °C isotherm (Fisher, 1978). Mather (1962) indicated that adults were commonly found near land or in less than 100 fathoms (183 m) depth and in waters deeper than 40 m (NOS, 1985). Kelley et al. (1990) reported that off Florida, blackfin tunas occur primarily along the shelf-slope edge with sporadic observations in the central Gulf of Mexico.

The NMFS longline database indicates that adult blackfin tunas are present throughout the western, central and the western half of the northern zone from January through March (Fig. 19). Densities in the eastern zone were generally low until June. One area of concentration early in the year appears to be the southwestern corner of the northern zone, which produced higher numbers of tunas during March. Another productive region in March was the area seaward of the 2000 m isobath in the central zone, which also produced higher catches (Fig. 19). By June, July, and August the catches of blackfin tunas were high throughout most of the study area with lowest landings in southern half of the eastern zone. By September, most tunas were taken over the slope water and landings diminished through the fall until in December, blackfin tunas were scattered through the western, central and northern zones with a few isolated catches in the eastern zone (Fig. 19).

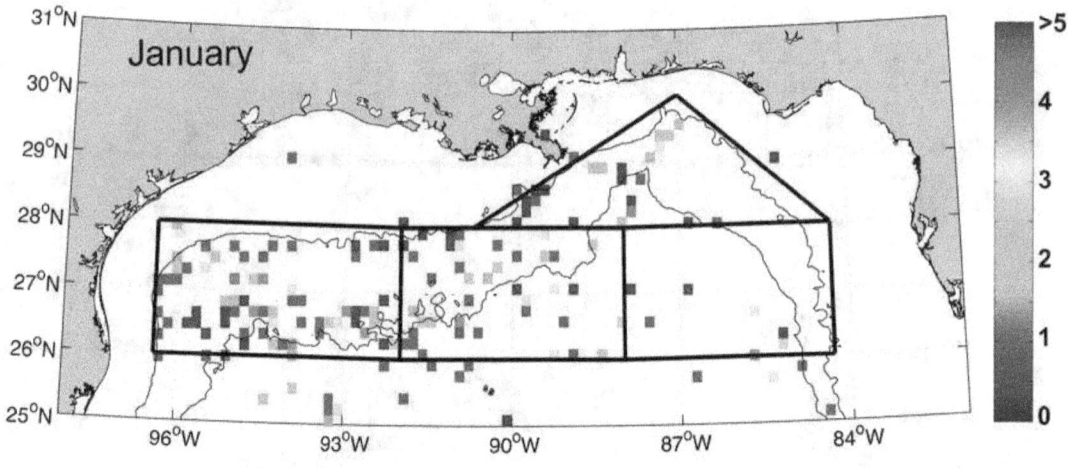

Figure 19.   Catch per unit effort of adult blackfin tuna from the commercial long-line fishery. Each square represents the mean catch-per-unit-effort (tuna per set) taken within a 10' x10' region from January through December over the period 1986-1999. Note that the colorbar was arbitrarily limited to a maximum CPUE of five. The maximum CPUEs were: January = 45; February = 32; March = 80; April = 50; May = 40; June = 54; July = 70; August = 40; September = 25; October = 25; November = 60; December = 100.

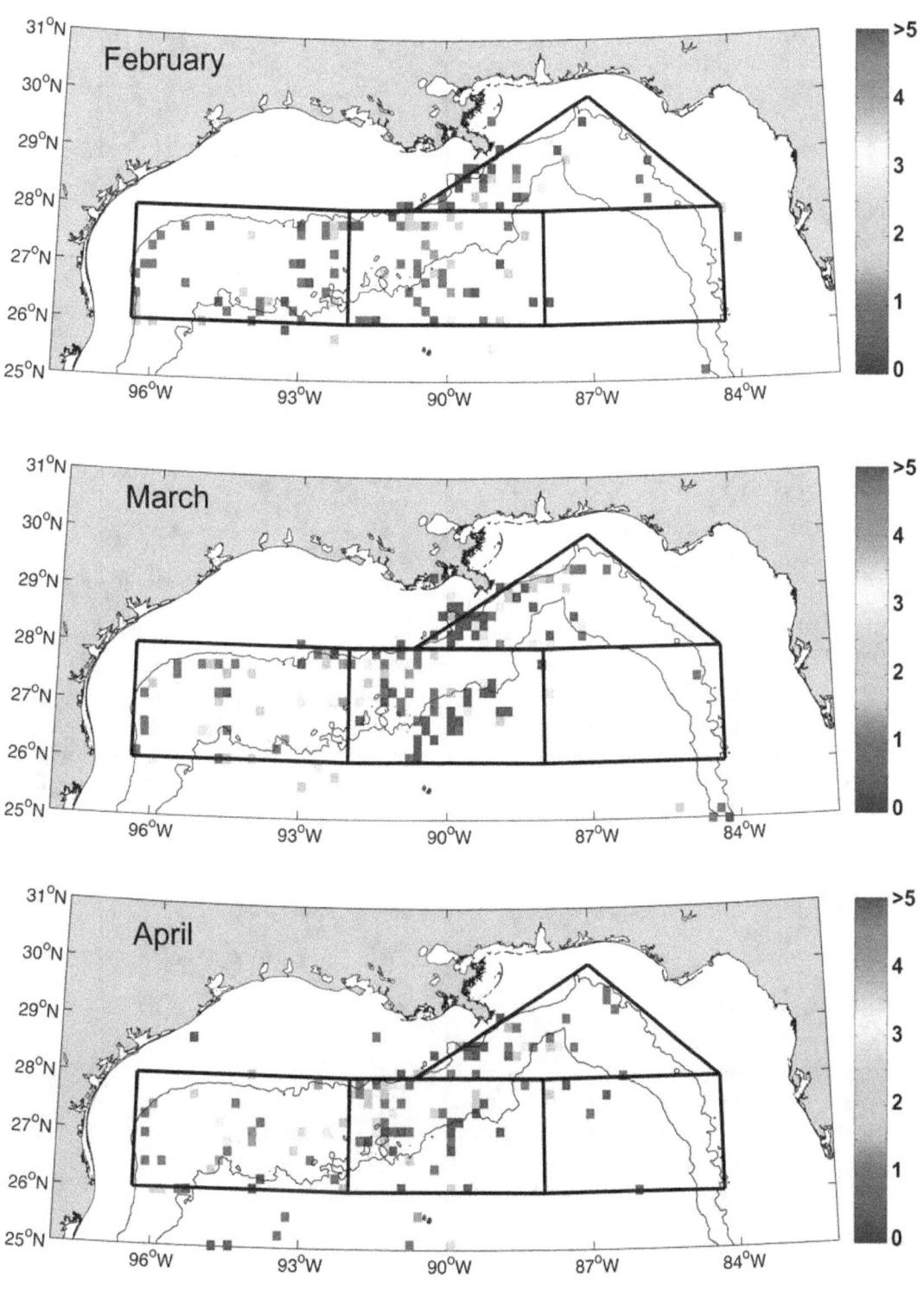

Figure 19. Catch per unit effort of adult blackfin tuna from the commercial long-line fishery. Each square represents the mean catch-per-unit-effort (tuna per set) taken within a 10' x10' region from January through December over the period 1986-1999. Note that the colorbar was arbitrarily limited to a maximum CPUE of five. The maximum CPUEs were: January = 45; February = 32; March = 80; April = 50; May = 40; June = 54; July = 70; August = 40; September = 25; October = 25; November = 60; December = 100. (continued)

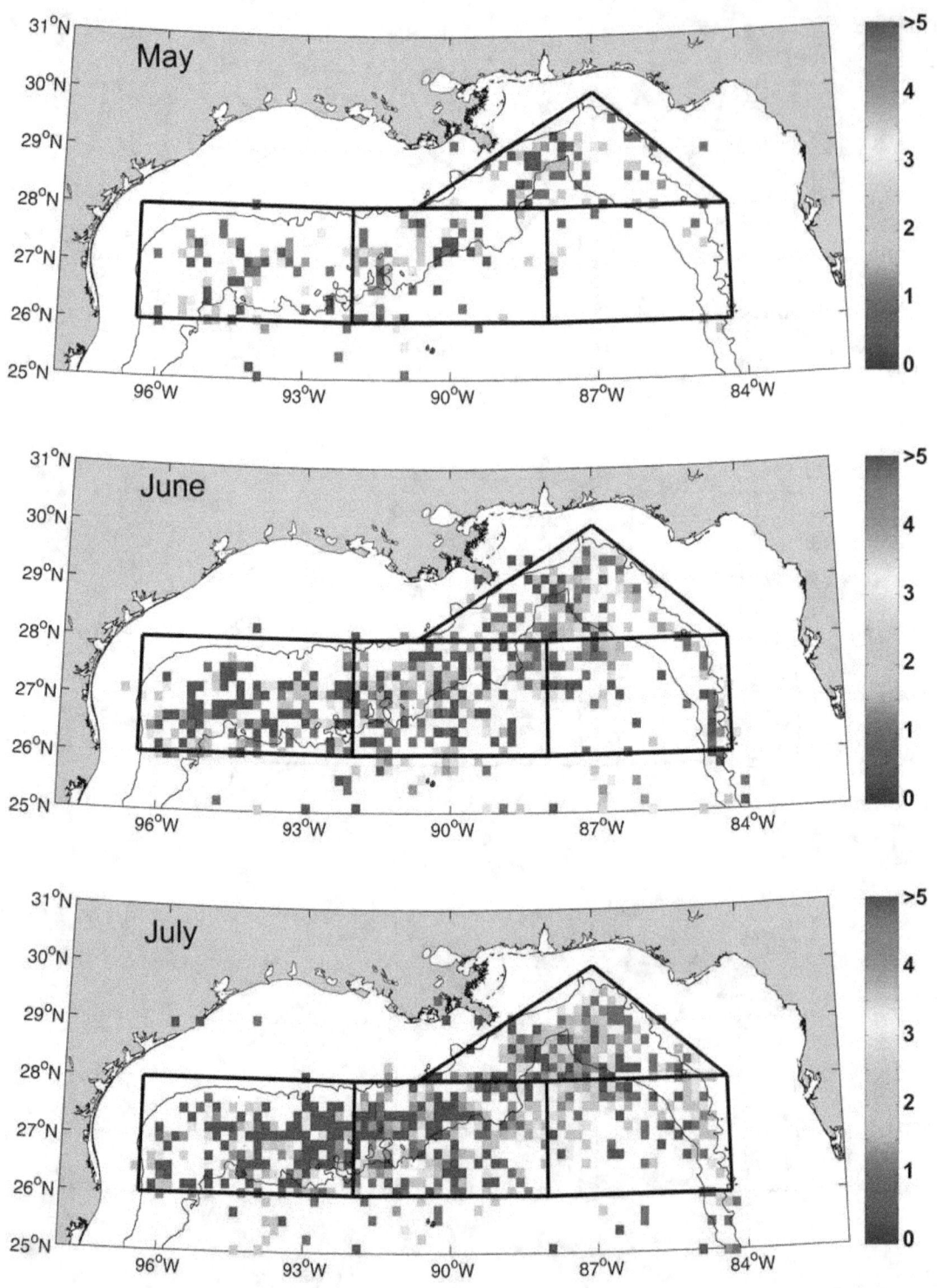

Figure 19. Catch per unit effort of adult blackfin tuna from the commercial long-line fishery. Each square represents the mean catch-per-unit-effort (tuna per set) taken within a 10' x10' region from January through December over the period 1986-1999. Note that the colorbar was arbitrarily limited to a maximum CPUE of five. The maximum CPUEs were: January = 45; February = 32; March = 80; April = 50; May = 40; June = 54; July = 70; August = 40; September = 25; October = 25; November = 60; December = 100. (continued)

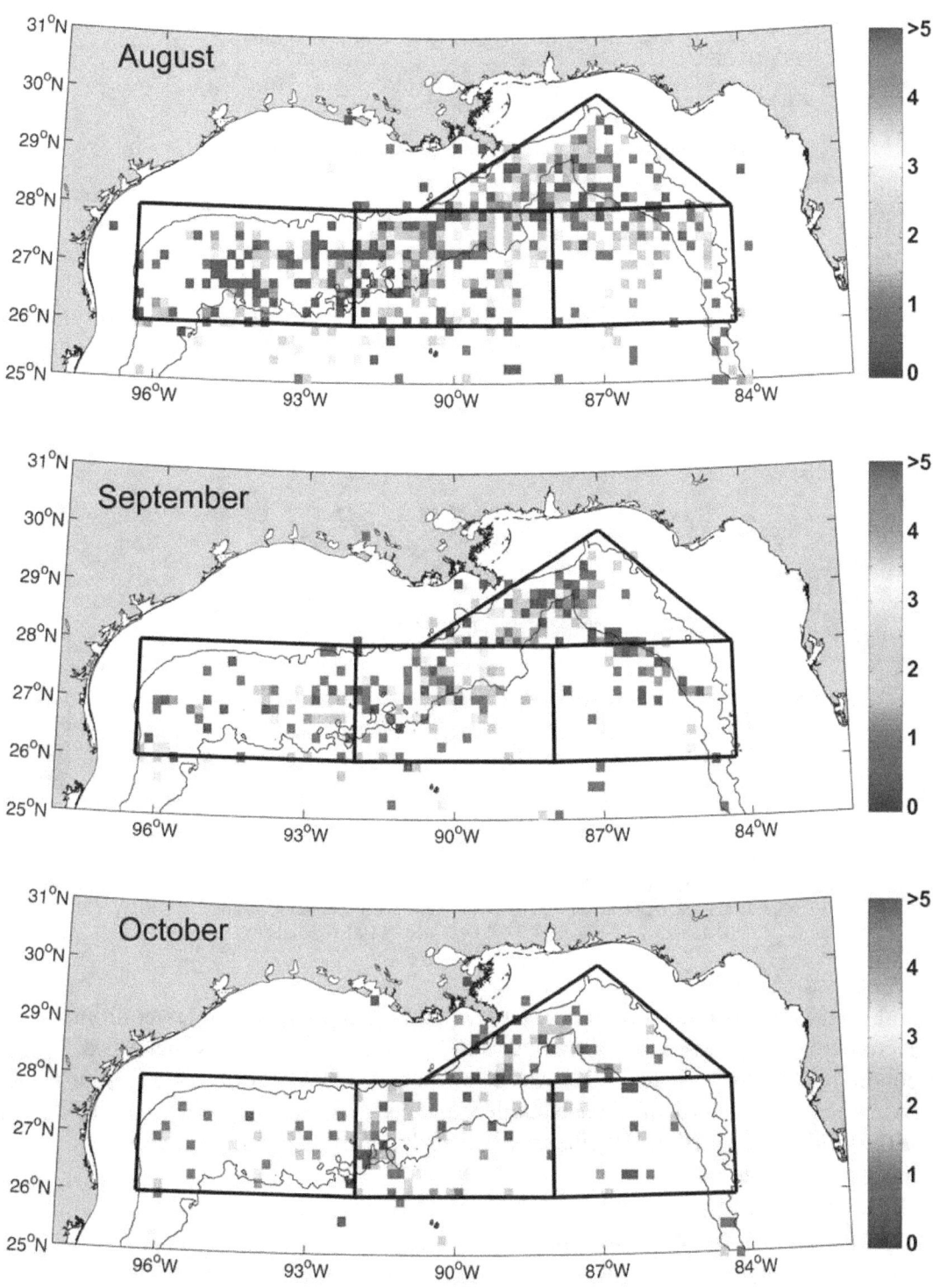

Figure 19. Catch per unit effort of adult blackfin tuna from the commercial long-line fishery. Each square represents the mean catch-per-unit-effort (tuna per set) taken within a 10' x10' region from January through December over the period 1986-1999. Note that the colorbar was arbitrarily limited to a maximum CPUE of five. The maximum CPUEs were: January = 45; February = 32; March = 80; April = 50; May = 40; June = 54; July = 70; August = 40; September = 25; October = 25; November = 60; December = 100. (continued)

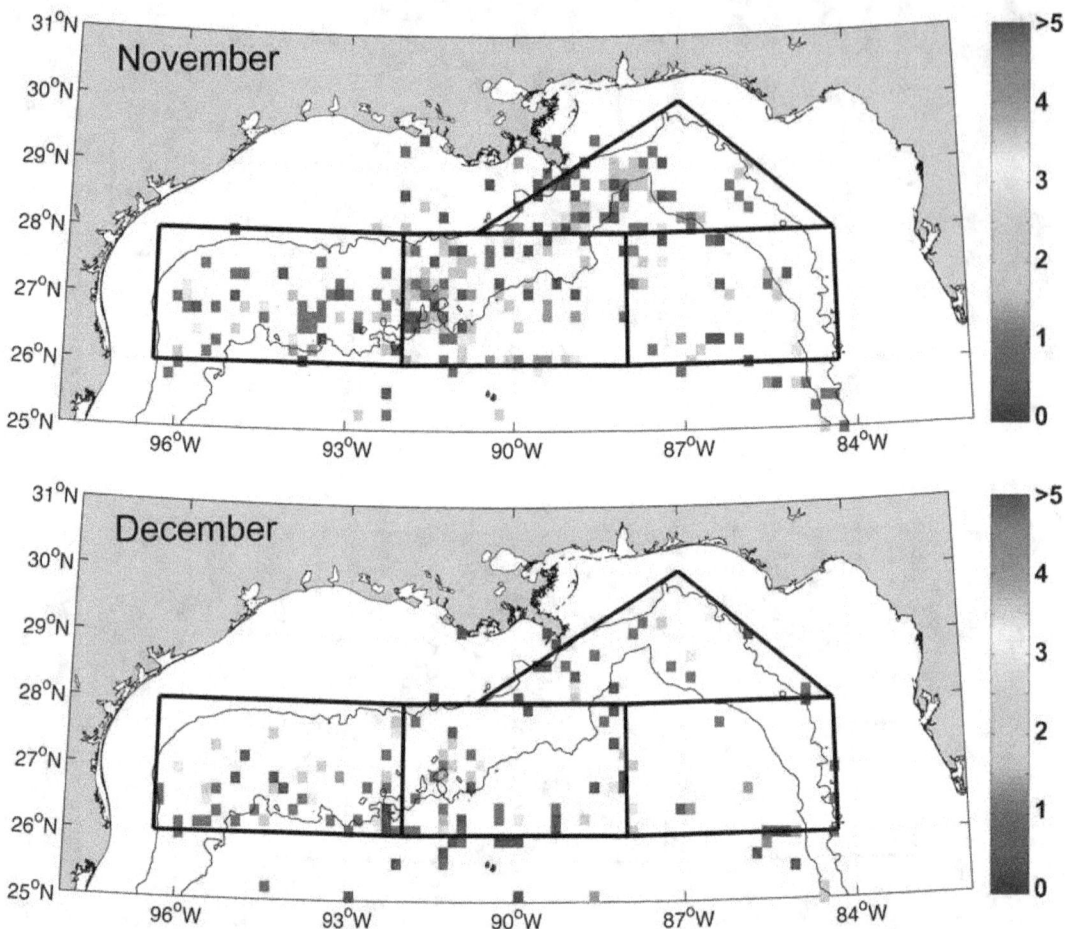

Figure 19.   Catch per unit effort of adult blackfin tuna from the commercial long-line fishery. Each square represents the mean catch-per-unit-effort (tuna per set) taken within a 10' x10' region from January through December over the period 1986-1999. Note that the colorbar was arbitrarily limited to a maximum CPUE of five. The maximum CPUEs were: January = 45; February = 32; March = 80; April = 50; May = 40; June = 54; July = 70; August = 40; September = 25; October = 25; November = 60; December = 100. (continued)

4.3.2 Reproduction

Ditty et al. (1988) reported the presence of larvae in the northern Gulf of Mexico from April through November with a peak in abundance from May through July. This suggests a corresponding spawning periodicity that was also reported by NOS (1985) in water throughout the oceanic region of the Gulf of Mexico. Collete and Nauen (1983) reported that blackfin tunas spawn off Florida from April to November with a peak during May and that they spawned in the Gulf from June to September.

4.3.3 Larval/Juvenile Distributions

It is difficult to distinguish the larvae of yellowfin and blackfin tunas (Grimes and Lang, 1992). Klawe and Shimada (1959) surveyed the distributions of scombrid larvae and juveniles off the Mississippi River plume and in other scattered locations in the Gulf of Mexico. While their data were not broken down by individual months, they found young blackfin tunas southeast of the Mississippi Delta between June and August and one individual on the southwestern edge of the western zone (Fig. 20). These young tunas were generally in waters over the 200-2000 m isobaths. Ditty et al. (1988) reviewed the temporal distributions of larvae in the Gulf and

58

indicated that while larvae may be present from April to November, peak abundances occurred from May through July.

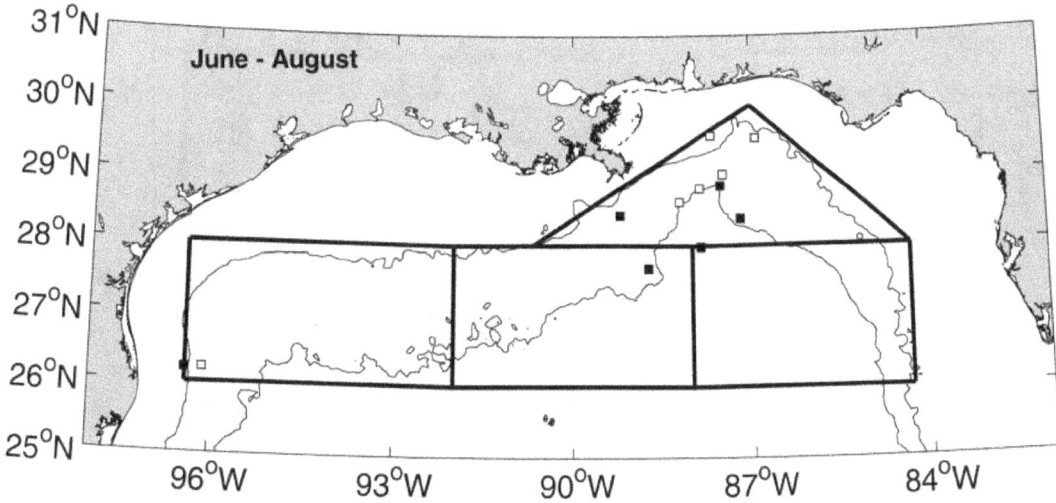

Figure 20. Distribution of larval and juvenile blackfin tuna during the period June 1954 to August 1956. Solid squares indicate blackfin tuna while open squares indicate individuals that were either blackfin or yellowfin tuna. Data digitized from Klawe and Shimada (1959) figure 7.

SEAMAP samples detected blackfin larvae and juveniles during April through August (Fig. 21). Few young tuna were detected in April and all were on the northern or western edge of the eastern zone seaward of the 2000 m isobath. By May, young blackfin were still sparse and present in the central regions of the central and eastern zone. A few individuals were present in the central waters of the eastern zone by June, however, the majority of young blackfin were found in the northern zone near the shelf-slope break and westward along the 200 m isobath. By July and August, young blackfin tunas were more abundant near the shelf-slope break in the western zone. This suggests a similar conveyor pattern to that observed for yellowfin tuna with young tunas produced near the Mississippi Delta prior to June being transported down-current to the west along the edge of the shelf in June, July, and August.

Vertically-stratified samples from a MOCNESS multinet sampler off Puerto Rico and the Virgin Islands indicated that small larvae (2-3.9 mm) were found in the upper 30 m of the water column, while larger larvae (>3.9 mm) were collected as deep as 60-80 m (Hare et al. 2001). The same study indicated that a diel vertical migration pattern by larvae was not evident, since similar numbers of larvae were collected in neuston nets during the day and night.

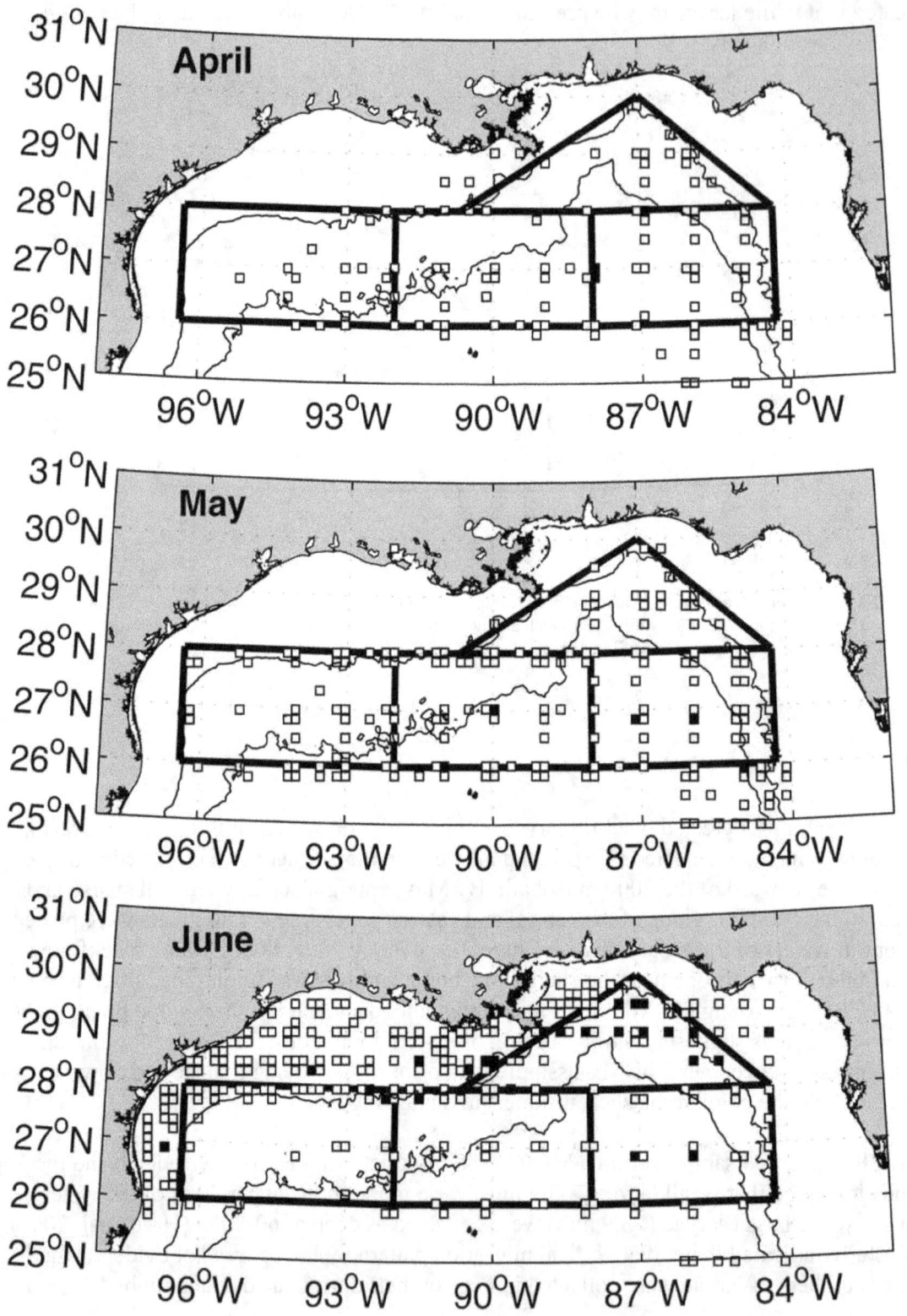

Figure 21. Presence (■) and absence (□) of blackfin tuna larvae in the study area during April, May, June, July, and August estimated from SEAMAP ichthyoplankton data.

60

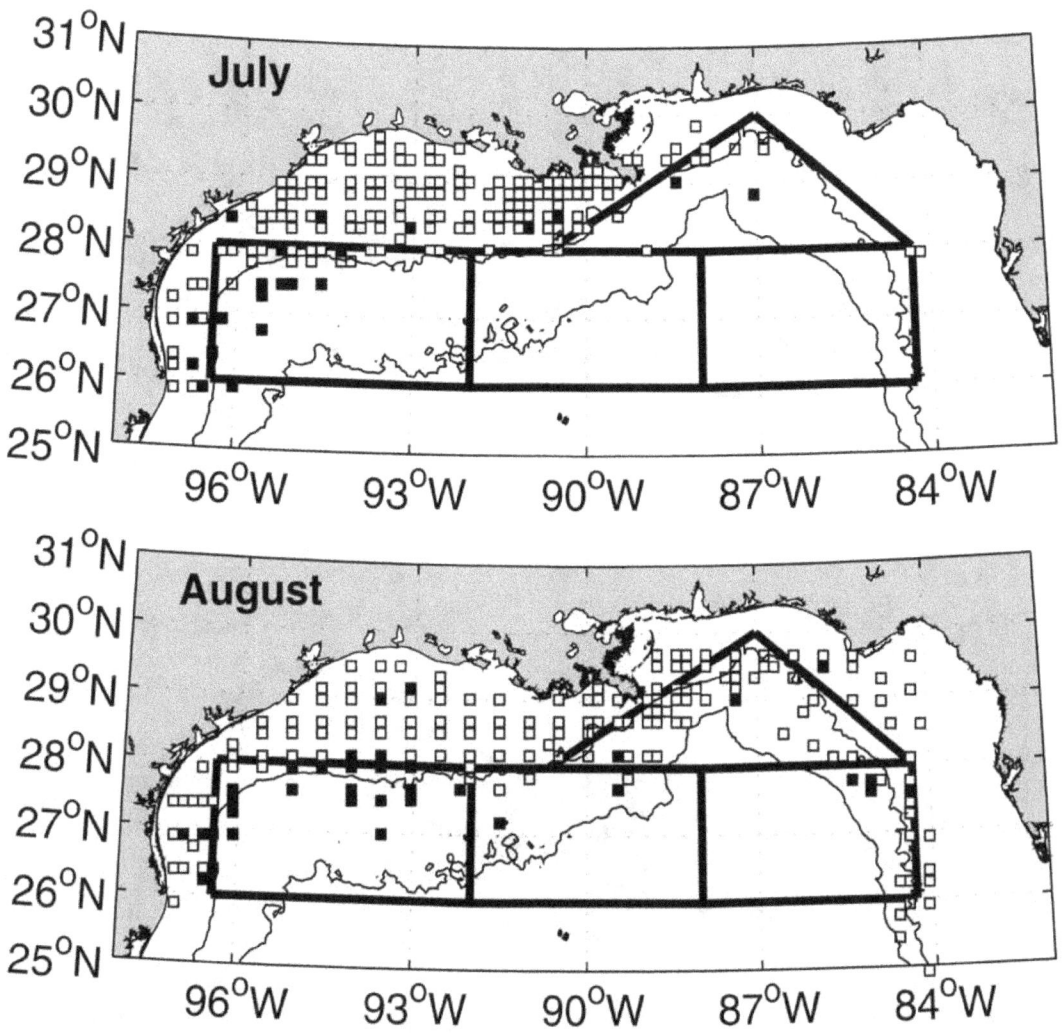

Figure 21.   Presence (■) and absence (□) of blackfin tuna larvae in the study area during April, May, June, July, and August estimated from SEAMAP ichthyoplankton data. (continued)

4.3.4 Predicted Adult Distributions

Adult blackfin tuna are predicted to be present throughout the western, most of the central, and the western half of the northern zone from January through April (Fig. 22). Though present in the eastern zone in January with the exception of the northeastern area landward of the 2000 m isobath (Fig. 22), they are largely absent from the eastern zone in February, March, and April. By May they occupy most of the western, central, and northern zones as well as the northern half of the eastern zone. In June, July, and August, blackfin adults are found throughout all four zones (Fig. 22). In September, adults are generally present seaward of the 200 m isobath in all four zones with absences in two pockets located in the south-central part of the eastern zone and at the southern junction of the eastern and central zone. From October through December, the distribution of adults becomes more fragmented. Fish are generally present in the majority of all four zones although, an increasing number of areas in the eastern and the eastern half of the northern zone are not predicted to contain fish (Fig. 22).

61

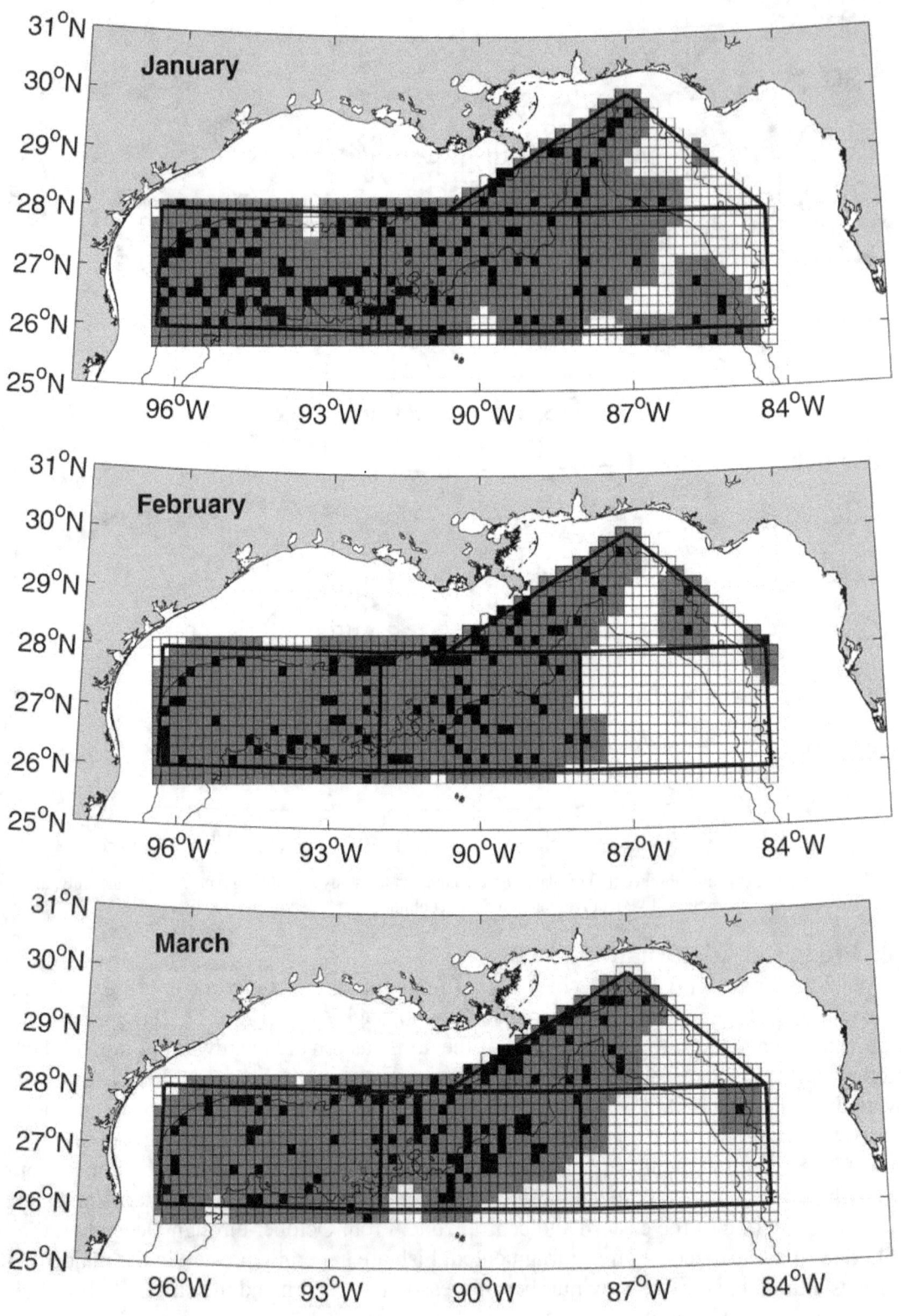

Figure 22. Predicted distributions of adult blackfin tuna in the study area from January through December. The presence of individuals in each grid cell is coded as: confirmed (■), reasonable inference (▨), or unreported (□).

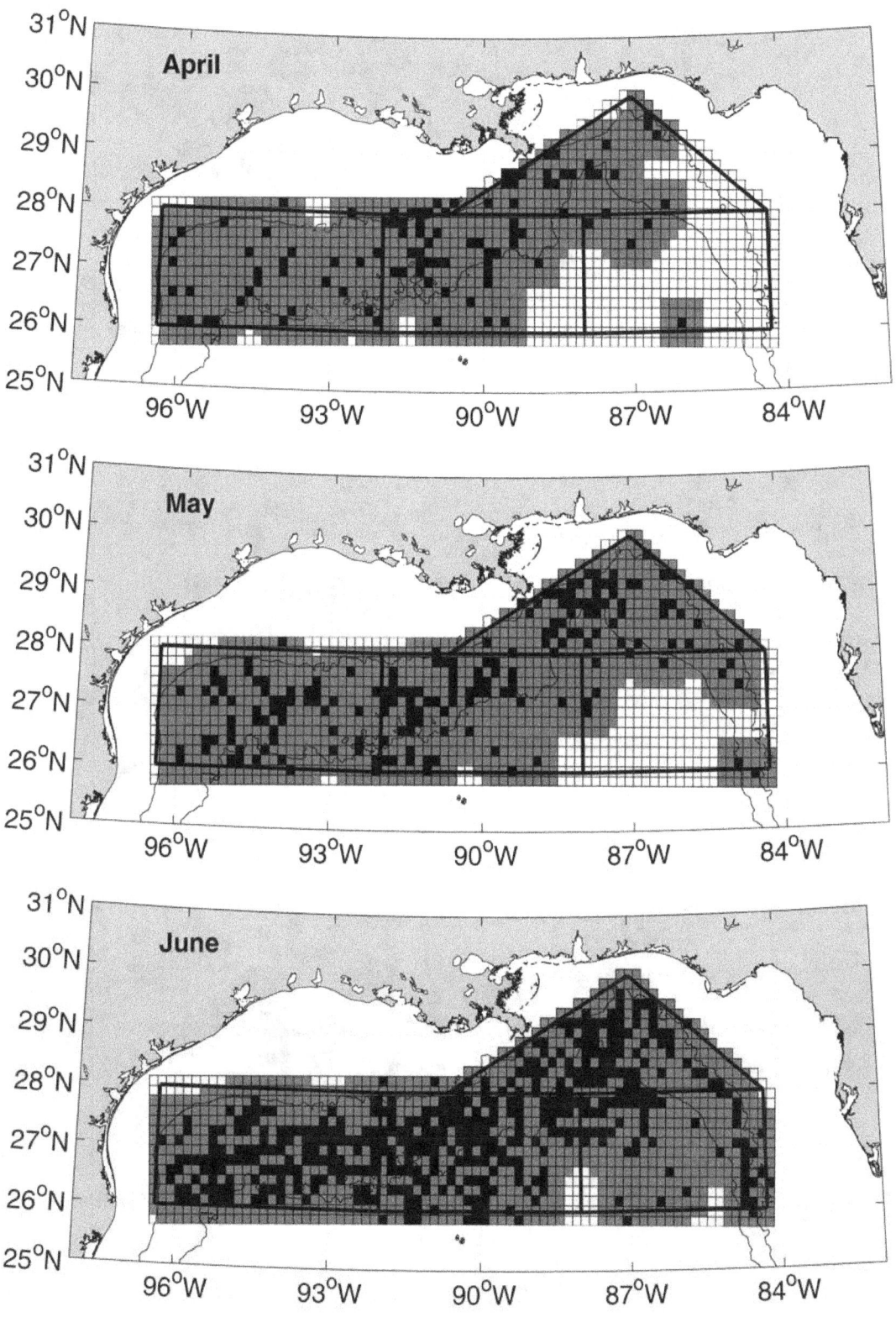

Figure 22. Predicted distributions of adult blackfin tuna in the study area from January through December. The presence of individuals in each grid cell is coded as: confirmed (■), reasonable inference (▨), or unreported (□). (continued)

63

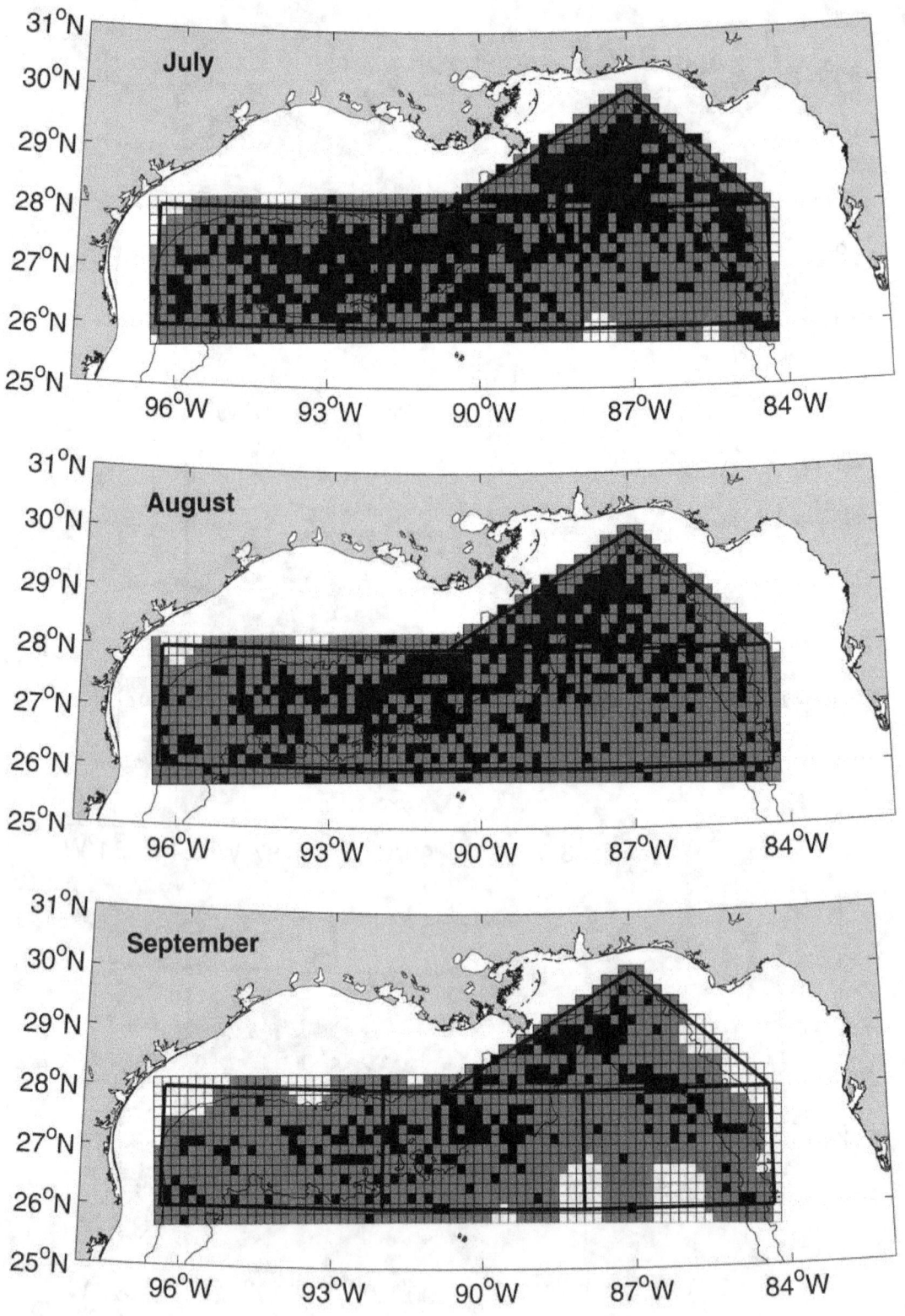

Figure 22. Predicted distributions of adult blackfin tuna in the study area from January through December. The presence of individuals in each grid cell is coded as: confirmed (■), reasonable inference (▨), or unreported (□). (continued)

64

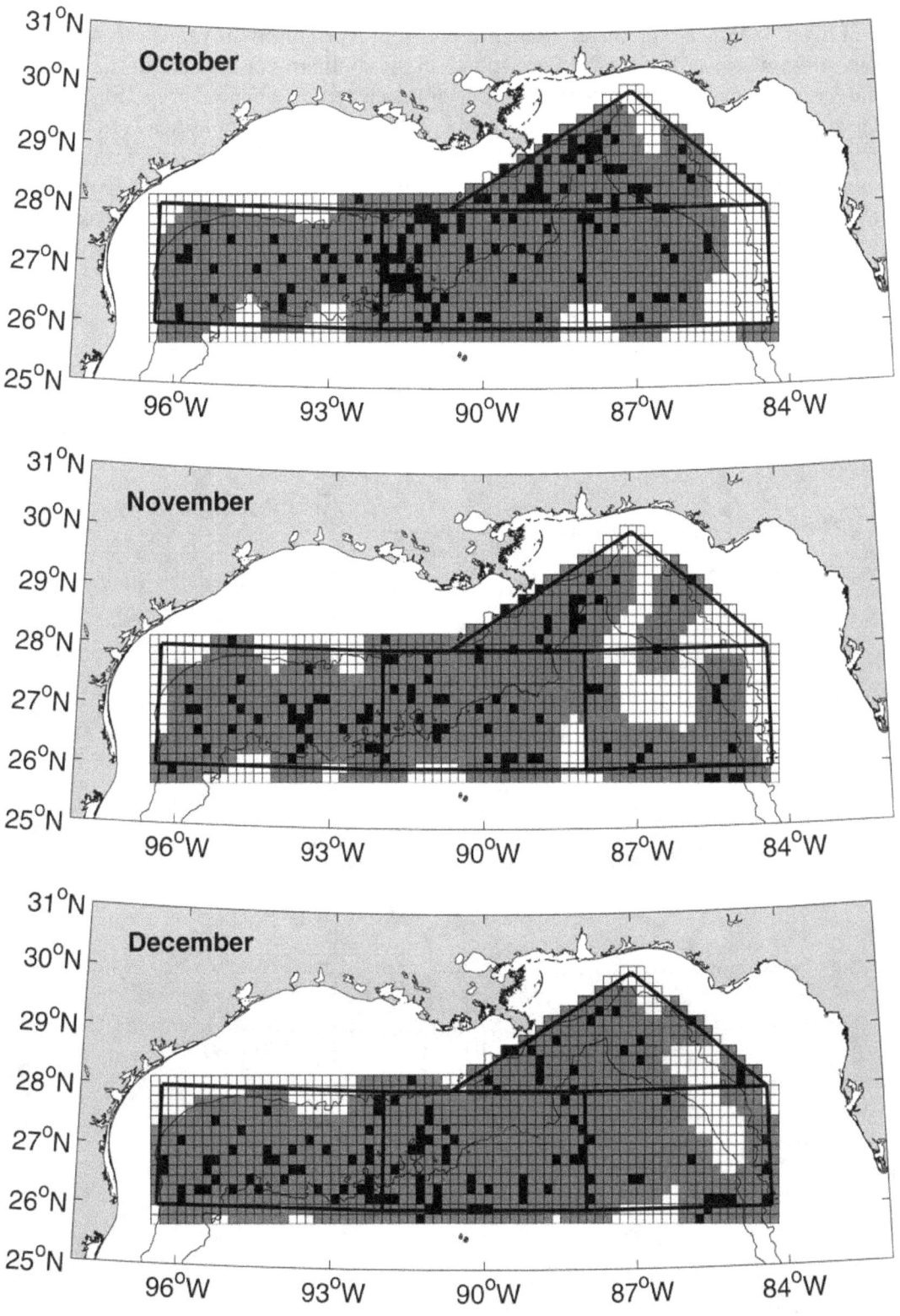

Figure 22. Predicted distributions of adult blackfin tuna in the study area from January through December. The presence of individuals in each grid cell is coded as: confirmed (■), reasonable inference (▦), or unreported (☐). (continued)

65

4.3.5 Predicted Larval/Juvenile Distributions

Larvae of blackfin tuna are predicted to be present from April through August. In April scattered larvae are present seaward of the 2000 m isobath in the southern-central area of the northern zone and along the central section of the border of the eastern and central zone (Fig. 23). By May, larvae are still scattered in the waters of the eastern and central zone seaward of the 2000 m isobath. A large zone of larvae occurs in the northern half of the northern zone in June and this patch extends along the 200 m isobath to the junction of the western and central zones (Fig. 23). Larvae produced in June appear to be advected westward along the shelf-slope break and accumulate (presumably as late larvae and juveniles) in the vicinity of the 200 m isobath in the waters of the western zone in July and August (Fig. 23).

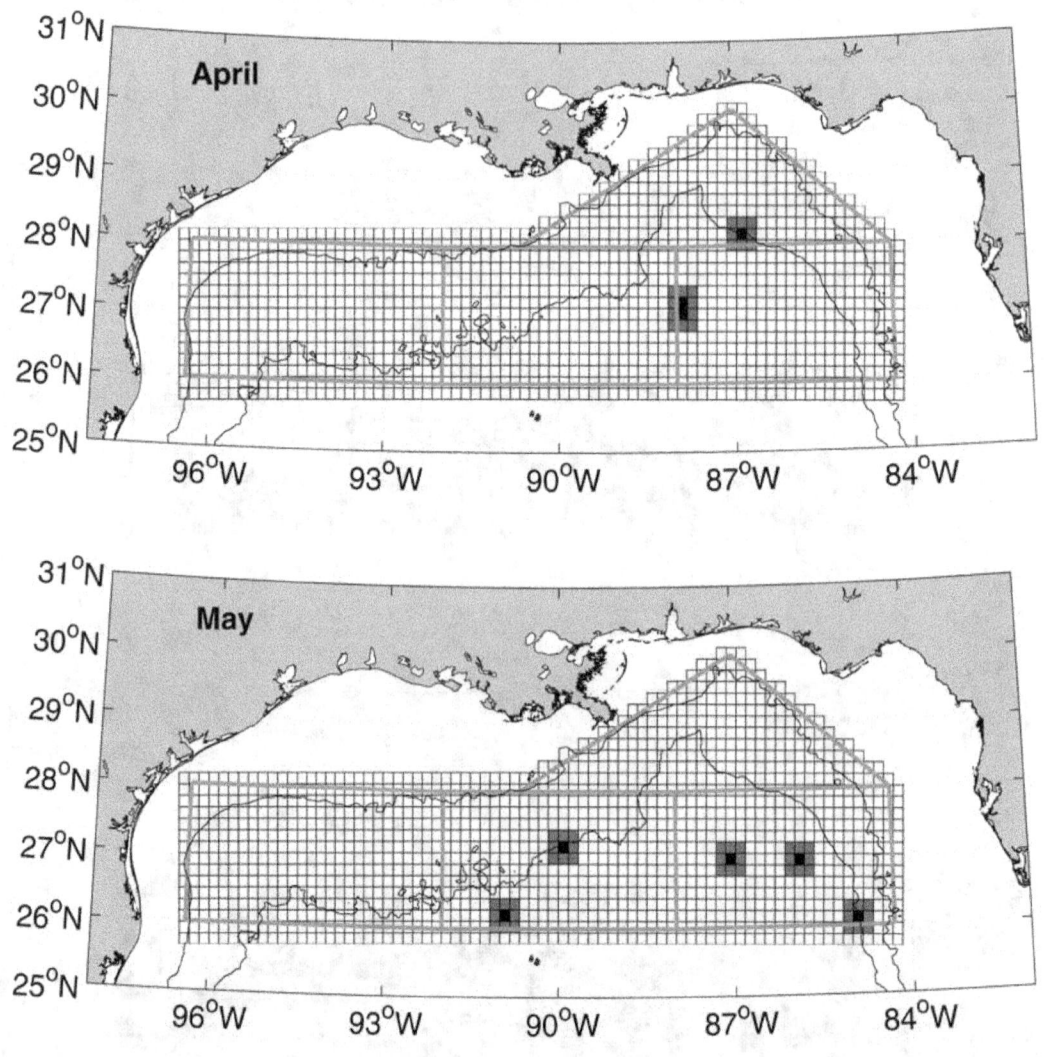

Figure 23. Predicted distributions of larval/juvenile blackfin tuna in the study area during April, May, June, July, and August. The presence of individuals in each grid cell is indicated as confirmed (■), reasonable inference (▦) or unreported (□).

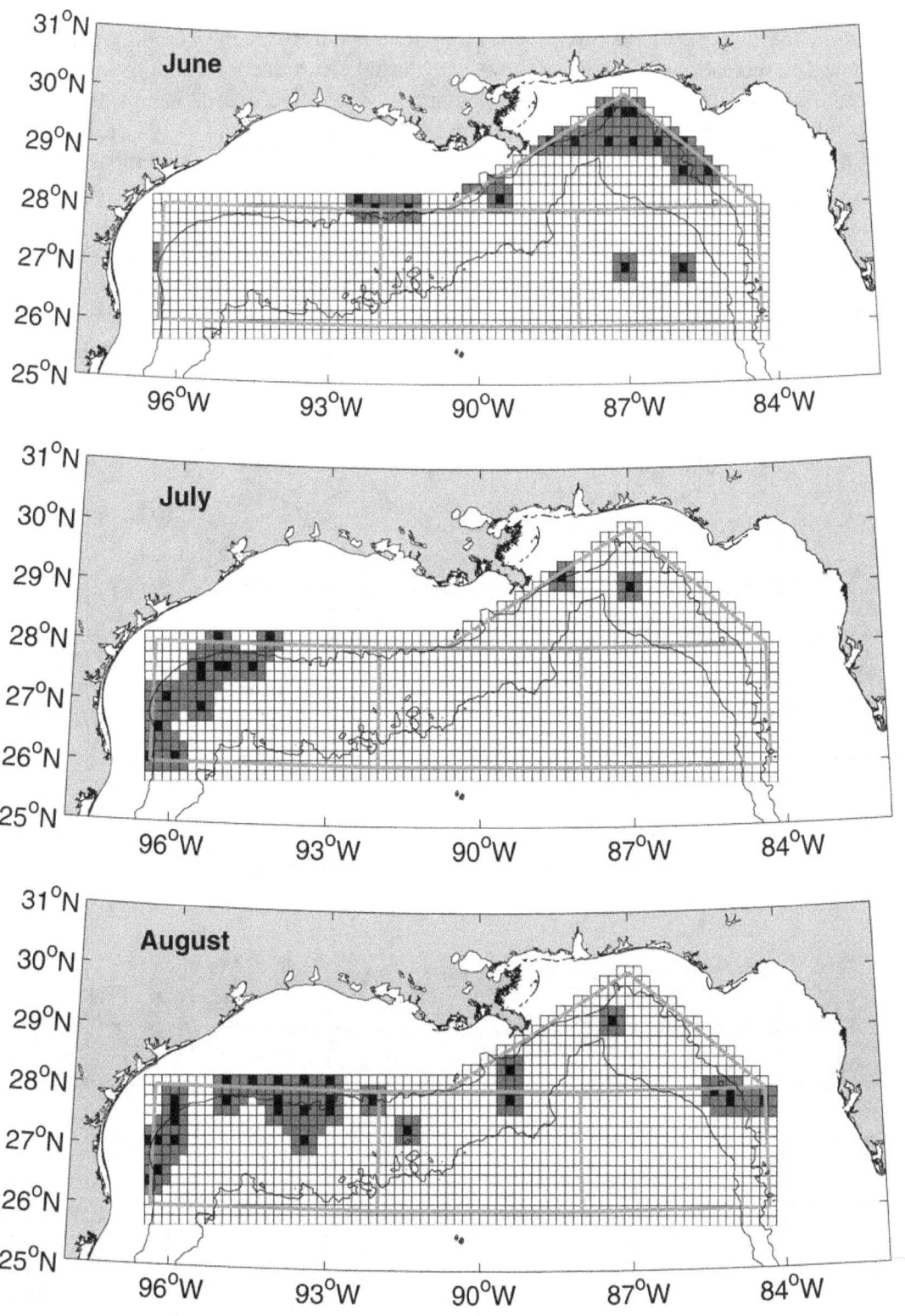

Figure 23. Predicted distributions of larval/juvenile blackfin tuna in the study area during April, May, June, July, and August. The presence of individuals in each grid cell is indicated as confirmed (■), reasonable inference (▨) or unreported (□). (continued)

67

4.4   Billfish: Blue Marlin and White Marlin

Billfish are exploited in a recreational fishery and are taken as bycatch in commercial long-lining operations. The recreational fishery is primarily a coastal effort due to limitations in range and endurance of the sportfishing vessels employed in the fishery. There are three primary recreational fishing zones (Fig. 24). Two of these zones (Panhandle and New Orleans) fall within the northern zone of the present study, while the Texas recreational zone encompasses the western and northwestern fringe of the present study's western zone (Fig. 24). Barry A. Vittor & Associates, Inc. (1985) used NMFS data to map the CPUE of billfishes in a region of the Gulf of Mexico east of the Mississippi Delta, south of Mississippi Sound and as far east as Pensacola, Florida. Their study area included parts of the New Orleans and Panhandle recreational fishing zones. The CPUE data were apparently a composite of all catches taken at different times of the year and various times of day. They indicate that the range of billfishes encompasses the northern half of the northern zone of the present study (Fig. 25).

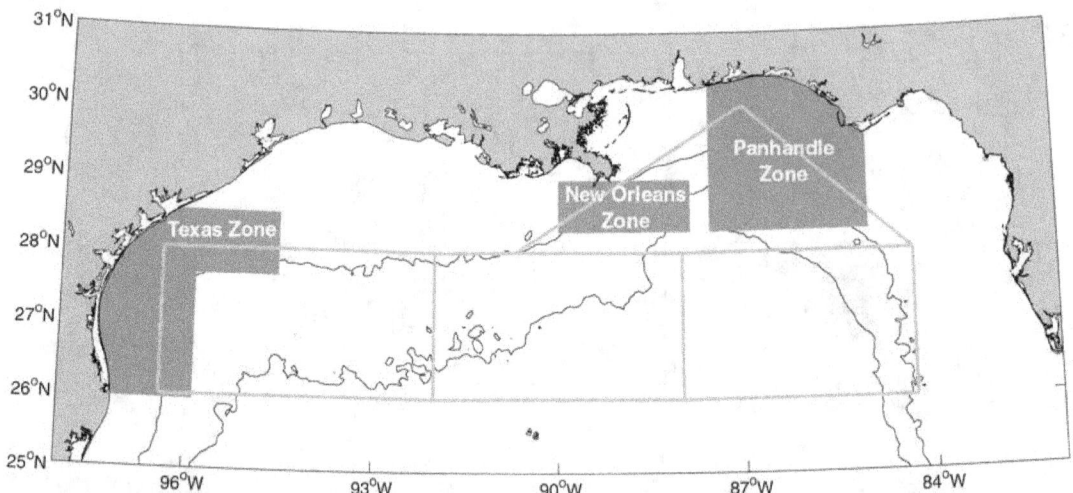

Figure 24. Primary areas where recreational fishing effort for billfishes is concentrated. After (Beardsley and Conser, 1981, Figure 3).

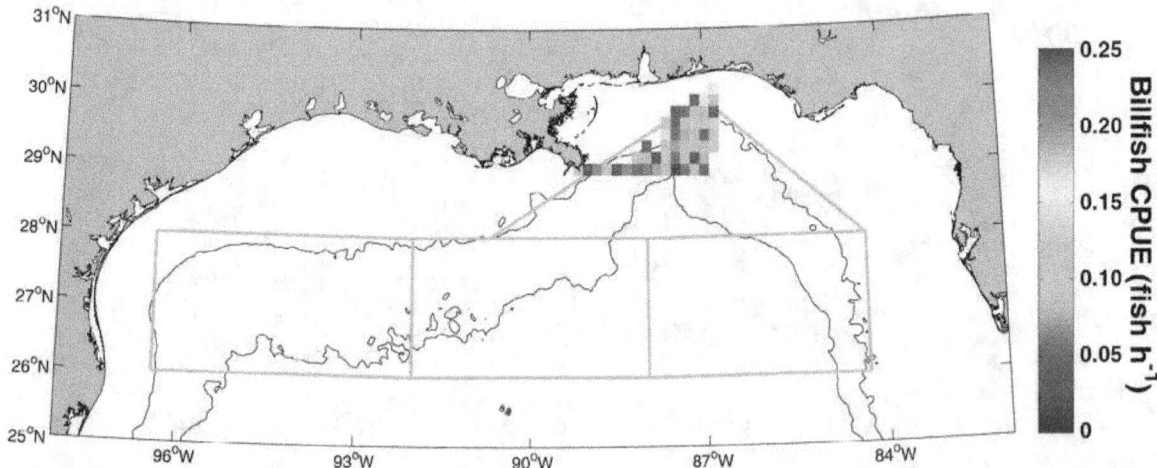

Figure 25. Catch per unit effort (CPUE) of billfishes expressed as fish per hour based on NMFS data. Modified from Figure 7.8 (Barry A. Vitor & Associates, Inc. (1985).

68

4.4.1   Blue Marlin (*Makaira nigricans*)
4.4.1.1   Adult Distributions

Billfishes, including blue marlin are abundant in the vicinity of the Loop Current and it has been speculated that the extent of their migration into the northeastern Gulf of Mexico is related to the northward extent of Loop Current in the region. Distributions of blue and white marlin may be inferred from an examination of the primary recreational fishing zones for billfishes. In the northern Gulf there are there primary recreational zones: Panhandle, New Orleans and Texas (Figure 24; Beardsley and Conser, 1981). An analysis of recreational tournament landings, dockside landings (reflecting non-competitive fishing) and Japanese longline catch data from 1971-1977 (Beardsley and Conser, 1981) suggest that blue marlin are present in the waters of this study's northern zone from March through September. The recreational tournament landings indicated that blue marlin are present nearer to the coast from July through September and appear in the offshore longline catch from March through September. This suggests that blue marlin are present in the deepwater areas frequented by the longliners and move closer inshore where they become vulnerable to recreational charter boats as the waters warm up in summer. This inshore migration is also suggested by NOS (1985) who report that blue marlin move into the northern Gulf of Mexico during spring and out during fall.

Acoustic telemetry of adult blue marlin off Hawaii indicated that fish swam deeper in the water during the day and near the surface at night (Holland et al. 1990). Telemetry studies by Block et al. (1992) off Hawaii indicated that blue marlin prefer warm 22-27 °C surface mixed layer and although fish are capable of diving through the thermocline, they seldom do so. Block et al. found considerable individual variation in the daily depth profiles of the six animals that they tagged. Most fish remained in the mixed layer and made rapid, brief descents at varying intervals. Over 50% of their time was spent within the upper 10 m of the water column (Block et al. 1992).

NMFS longline bycatch data for blue marlin reveal that this species is present in all four study zones during January although no fish were taken north of the northern-most extent of the 2000 m isobath in the northern zone and most were present close to, or south of the 2000 m isobath in the central and western zones (Fig. 26). During February, March, and April, CPUEs remained low and fish moved north while generally remaining seaward of the 200 m isobath (Fig. 26). In May and June, the numbers of fish taken increased and blue marlin occurred throughout all four zones with concentrations in the western, central, and the western half of northern zone (Fig. 26). This pattern persisted during June, July, and August. From September through December, the abundance of blue marlin in the bycatch diminished as fish moved towards the southwest and presumably emigrated from the cooling waters of the northern Gulf (Fig. 26).

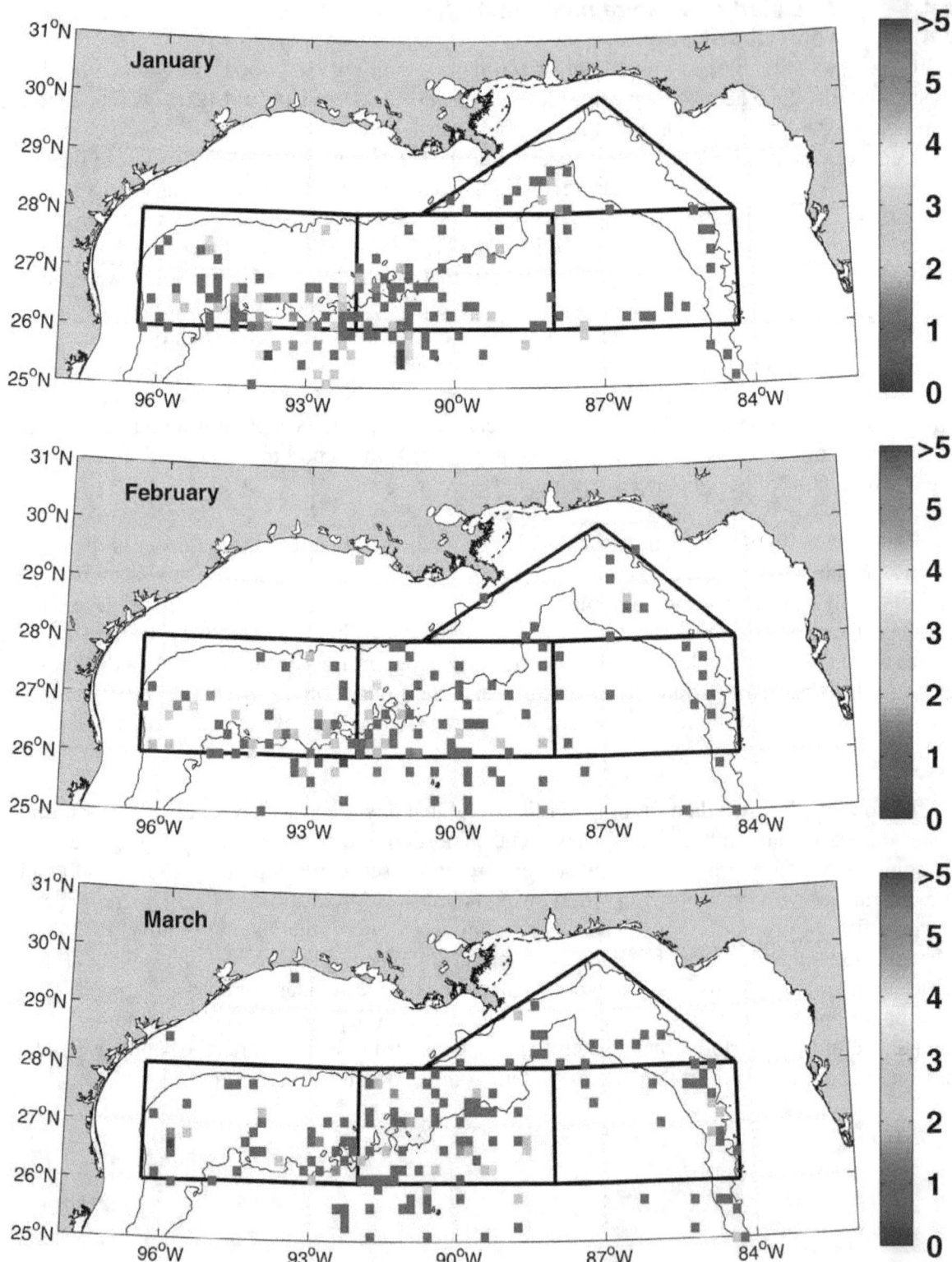

Figure 26. Catch per unit effort of adult blue marlin from the commercial long-line fishery from January through December. Each square represents the mean catch-per-unit-effort (blue marlin per set) taken within a 10' x10' region over the period 1986-1999. Note that the colorbar was arbitrarily limited to a maximum CPUE of five. The maximum CPUEs were: January = 10; February = 7; March = 8; April = 18.5; May = 12; June = 50; July = 27; August = 43; September = 23; October = 13.5; November = 11; December = 17.

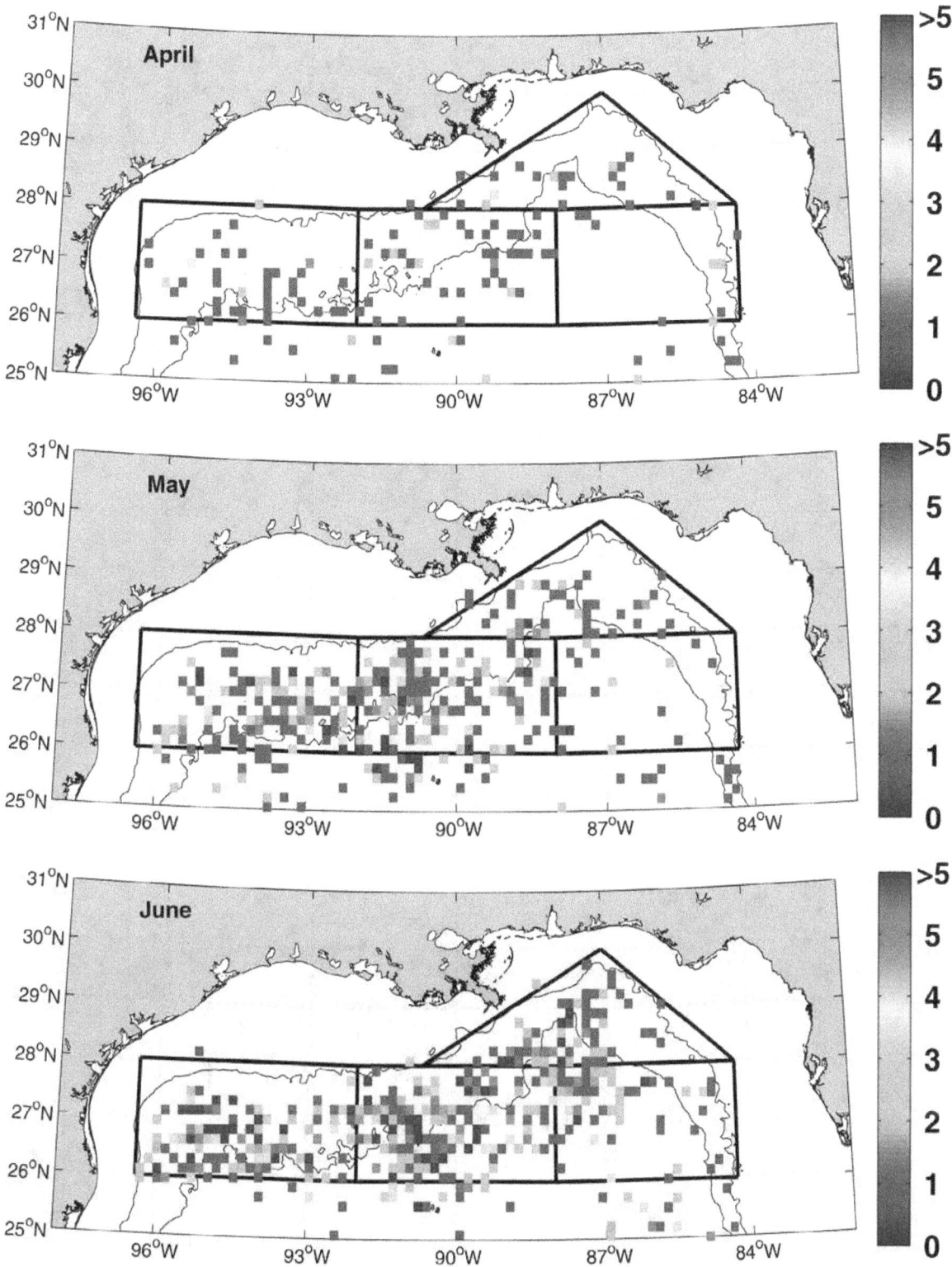

Figure 26. Catch per unit effort of adult blue marlin from the commercial long-line fishery from January through December. Each square represents the mean catch-per-unit-effort (blue marlin per set) taken within a 10' x10' region over the period 1986-1999. Note that the colorbar was arbitrarily limited to a maximum CPUE of five. The maximum CPUEs were: January = 10; February = 7; March = 8; April = 18.5; May = 12; June = 50; July = 27; August = 43; September = 23; October = 13.5; November = 11; December = 17. (continued)

71

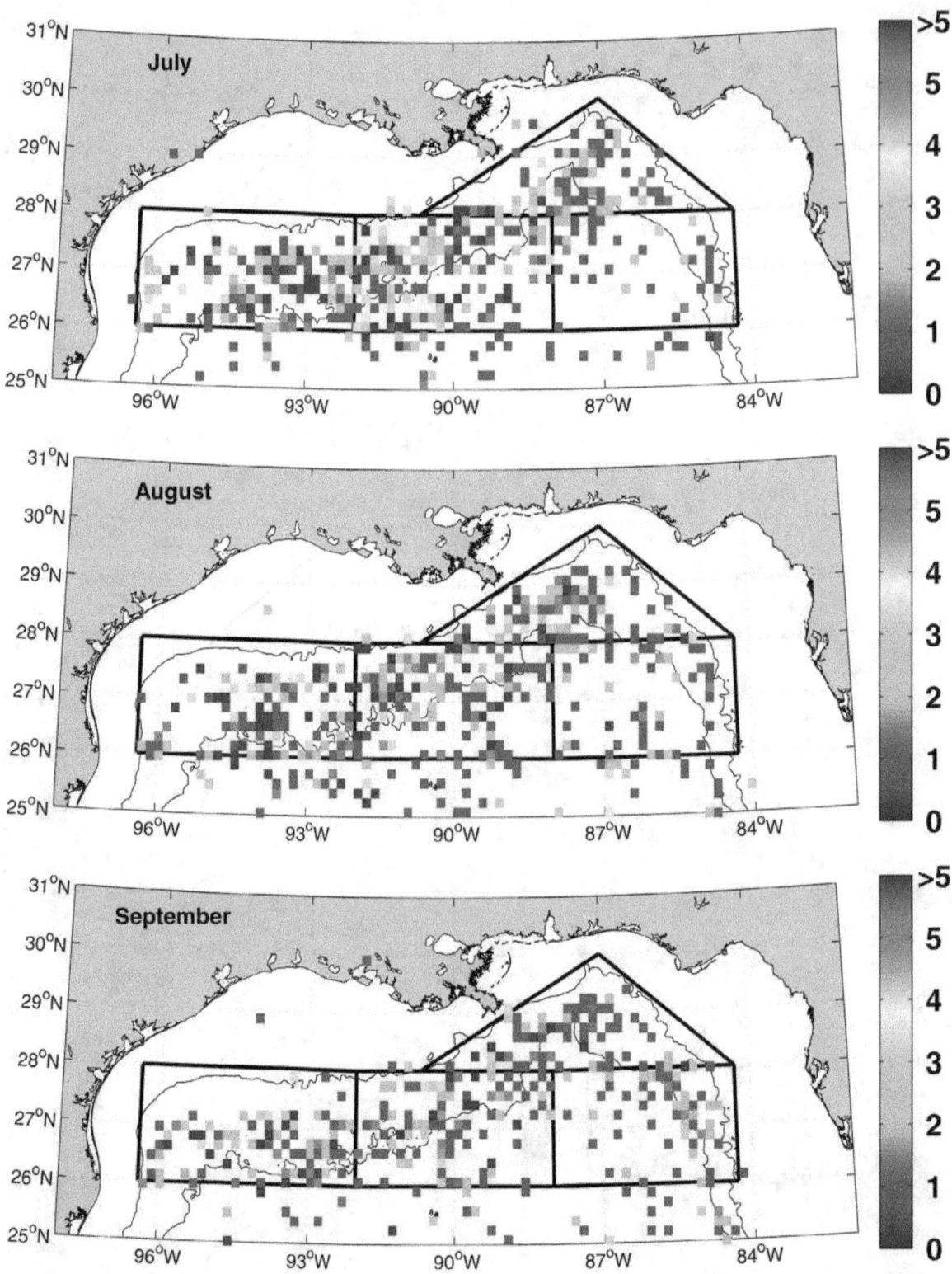

Figure 26. Catch per unit effort of adult blue marlin from the commercial long-line fishery from January through December. Each square represents the mean catch-per-unit-effort (blue marlin per set) taken within a 10' x10' region over the period 1986-1999. Note that the colorbar was arbitrarily limited to a maximum CPUE of five. The maximum CPUEs were: January = 10; February = 7; March = 8; April = 18.5; May = 12; June = 50; July = 27; August = 43; September = 23; October = 13.5; November = 11; December = 17. (continued)

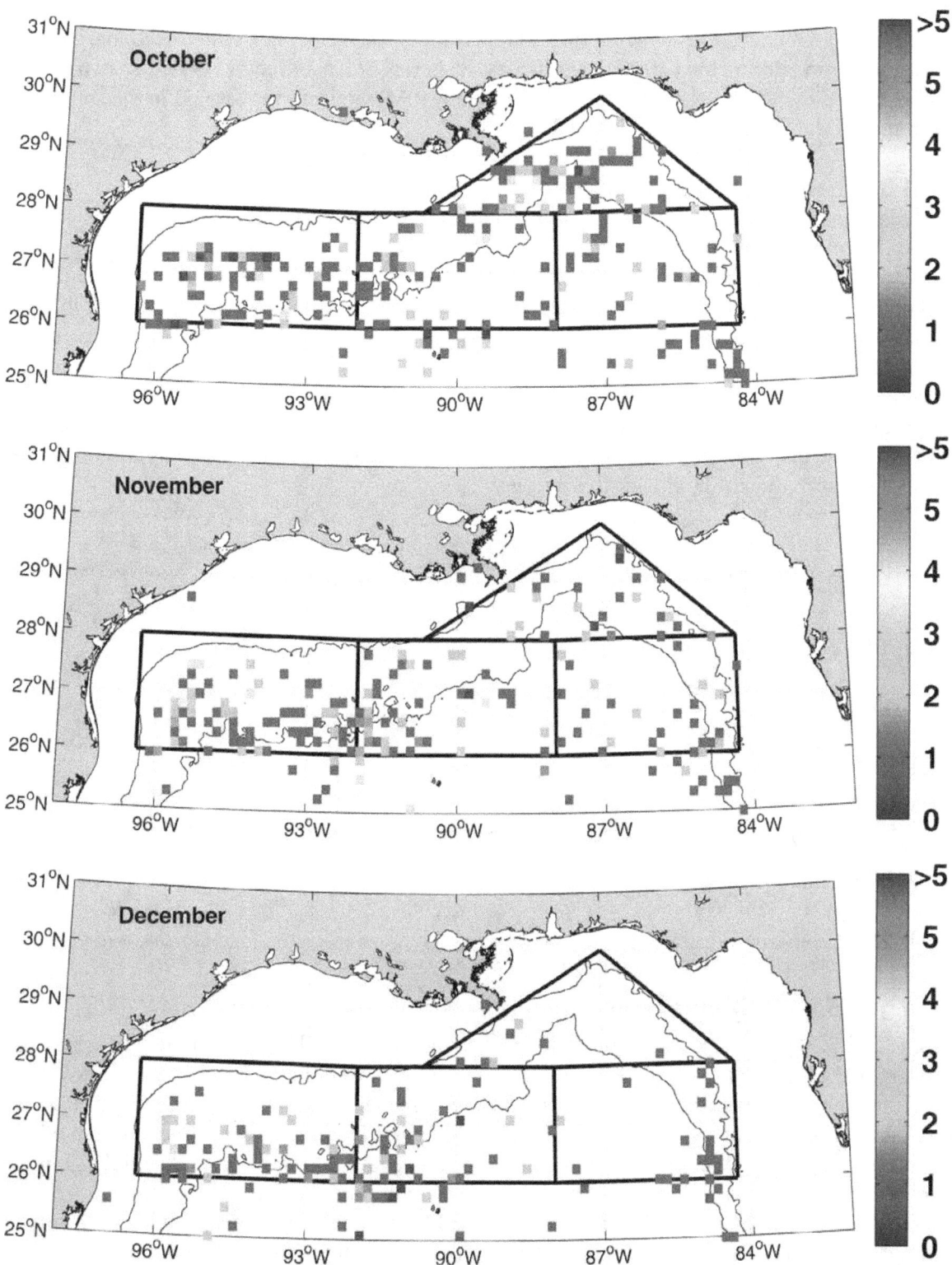

Figure 26. Catch per unit effort of adult blue marlin from the commercial long-line fishery from January through December. Each square represents the mean catch-per-unit-effort (blue marlin per set) taken within a 10' x10' region over the period 1986-1999. Note that the colorbar was arbitrarily limited to a maximum CPUE of five. The maximum CPUEs were: January = 10; February = 7; March = 8; April = 18.5; May = 12; June = 50; July = 27; August = 43; September = 23; October = 13.5; November = 11; December = 17. (continued)

4.4.1.2 Reproduction

Blue marlin are believed to spawn during summer in the Caribbean Sea with a secondary spawning period during the early fall (Southeast Fisheries Science Center, 1992). Spawning occurs during the warm months in the northern Gulf of Mexico and year-round in the southern Gulf and off the Florida Keys (NOS, 1985).

4.4.1.3 Larval/Juvenile Distributions

Little is known of the distributions of larval and juvenile blue marlin in the Gulf of Mexico. Since spawning occurs offshore in oceanic waters, it has been suggested that the nursery areas are throughout the distributional range of the adult fishes (NOS, 1985). The SEAMAP dataset contained only two records of blue marlin larvae. One was collected seaward of the 200 m isobath in the northern zone during June and the other at the southern boundary of the eastern and central zones in May (Fig. 27). On the basis of such sparse distributional data, little can be directly inferred about where larvae occur. The presence of a larval blue marlin in May indicates that spawning occurs as early as May and extends through the summer according to NOS (1985). Therefore, the distribution of adults from May through August may provide an approximation of the likely distribution of larvae for this species.

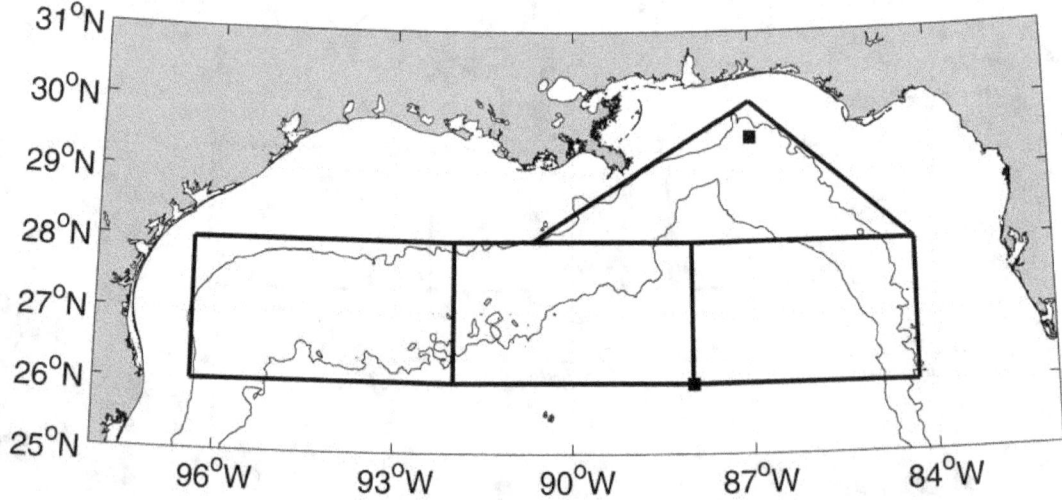

Figure 27. Locations of two blue marlin larvae collected during SEAMAP sampling. The individual in the northern zone was collected during June and the other larva was collected during May.

4.4.1.4 Predicted Adult Distributions

From January through March, the predicted distributions of adult blue marlin include most of the western and central zones, as well as large patches in the northern and eastern zones. Few confirmed observations occurred shoaler than the 200 m isobath and by March, most fish were predicted to be south of 29 °N (Fig. 28). In April, blue marlin move closer to the northern edge of the northern zone and occupy most of the western and central zones. A large area of the central part of the eastern zone is devoid of fish. During May and June the distribution expands to encompass the majority of all four zones (Fig. 28). This complete coverage persists from July, through November. By December, the distributions tend to shift south in the western and central zones. At this time, blue marlin are largely absent from the northern and central region of the northern zone and from the central region of the eastern zone (Fig. 28).

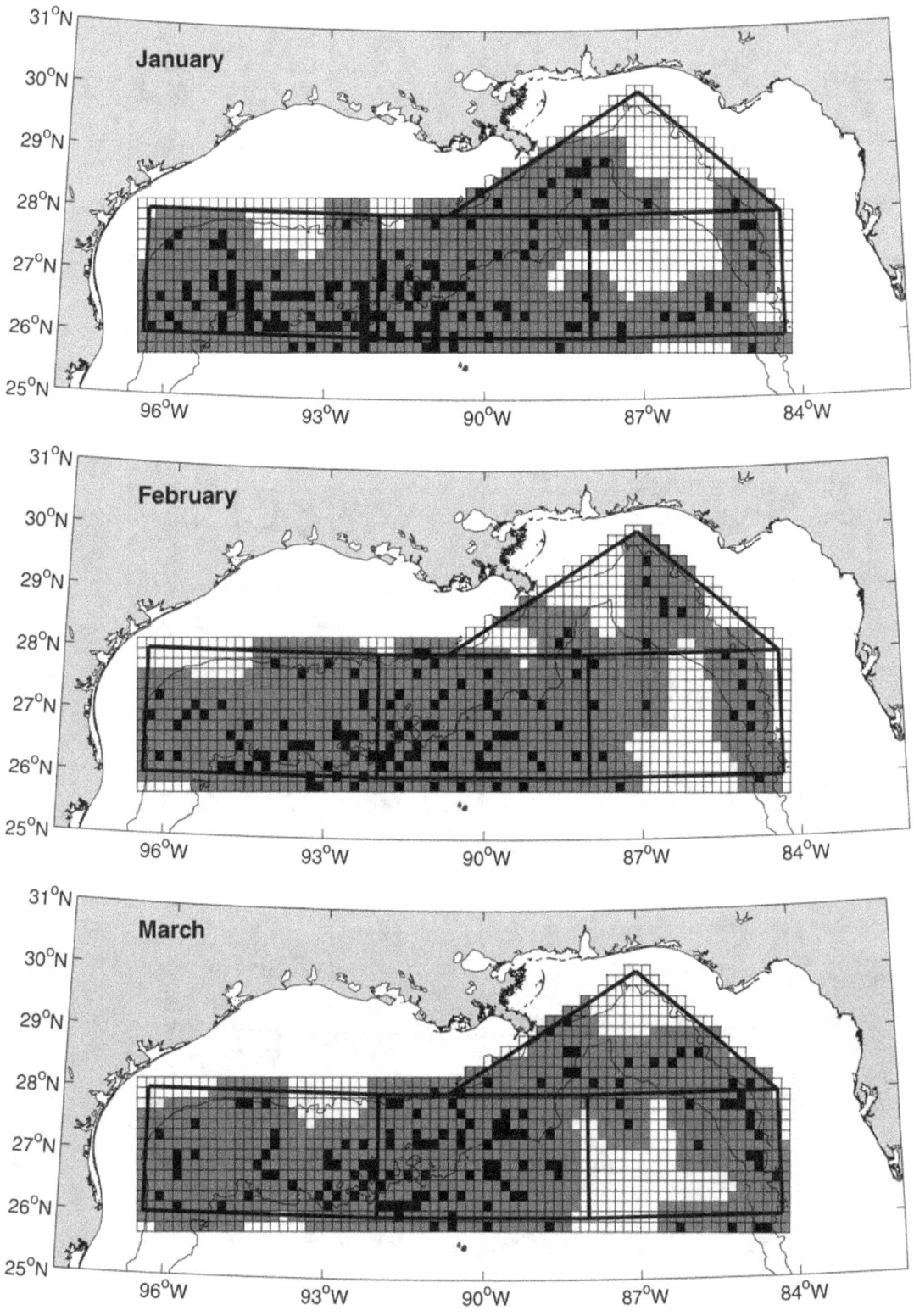

Figure 28. Predicted distributions of adult blue marlin in the study area from January through December. The presence of individuals in each grid cell is coded as: confirmed (■), reasonable inference (▨), or unreported (□).

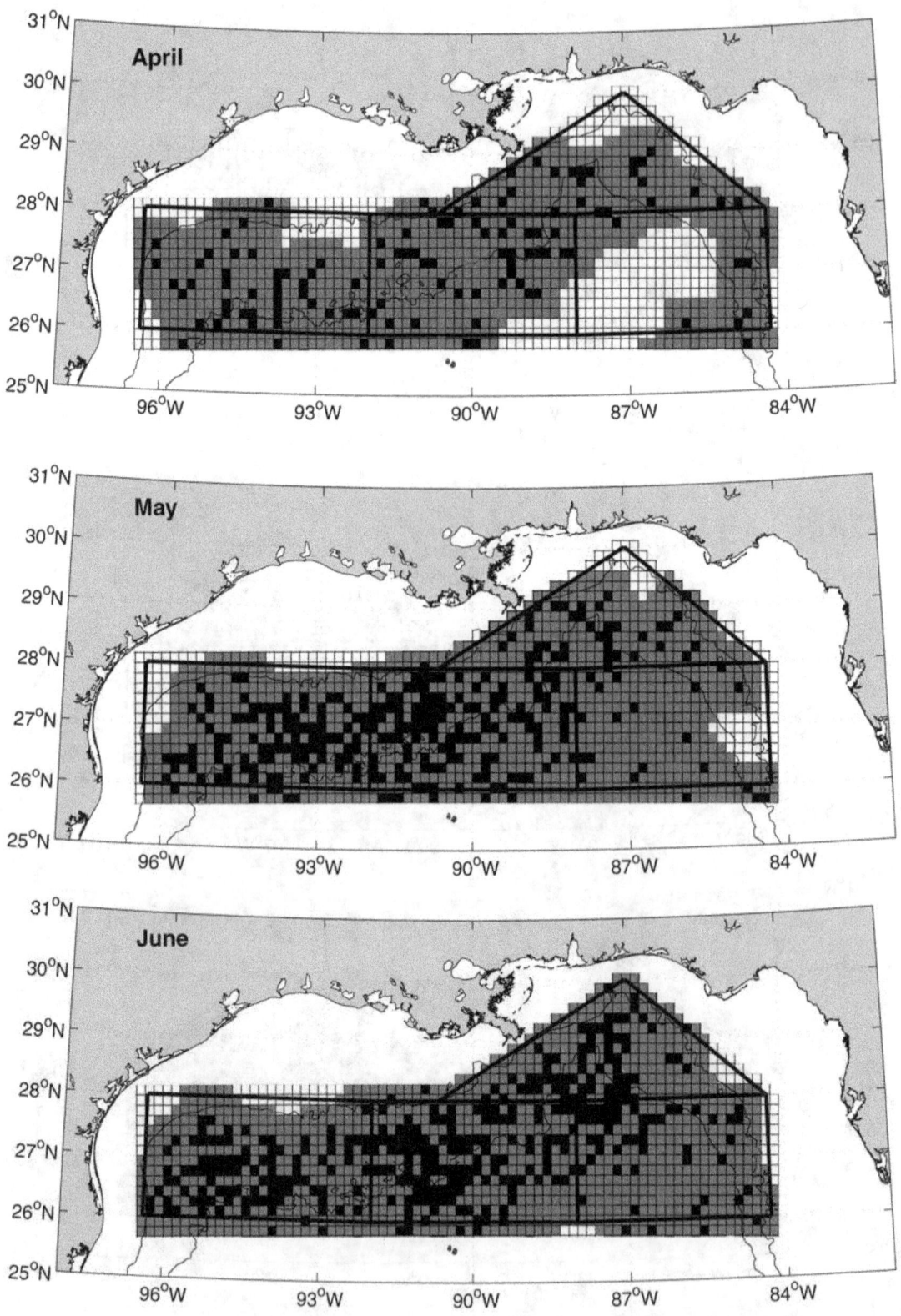

Figure 28. Predicted distributions of adult blue marlin in the study area from January through December. The presence of individuals in each grid cell is coded as: confirmed (■), reasonable inference (▨), or unreported (□). (continued)

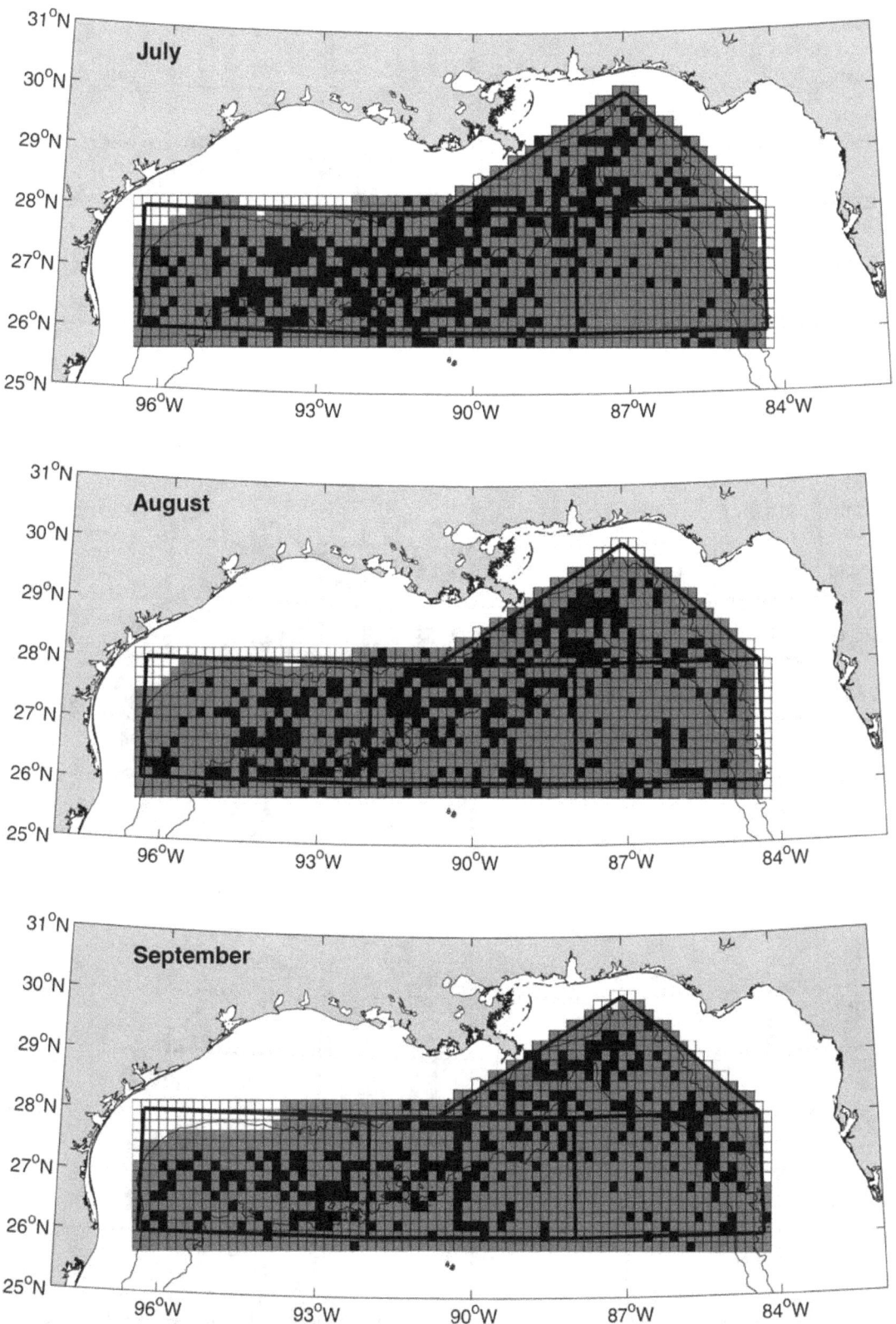

Figure 28. Predicted distributions of adult blue marlin in the study area from January through December. The presence of individuals in each grid cell is coded as: confirmed (■), reasonable inference (▨), or unreported (□). (continued)

77

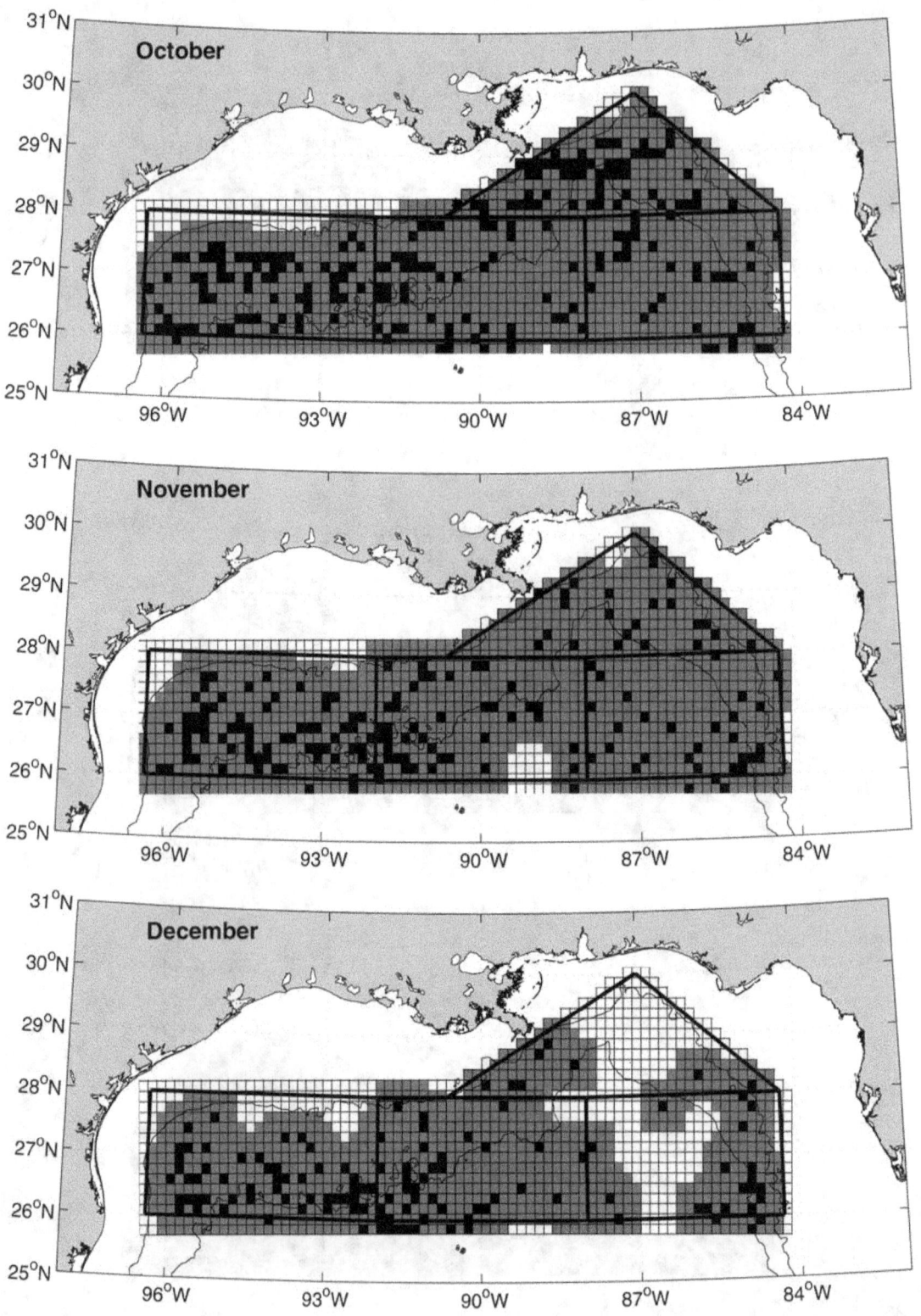

Figure 28. Predicted distributions of adult blue marlin in the study area from January through December. The presence of individuals in each grid cell is coded as: confirmed (■), reasonable inference (▨), or unreported (□). (continued)

78

4.4.1.5 Predicted Larval/Juvenile Distributions

With only two confirmed larval samples in the SEAMAP database, it is unrealistic to use these data to predict the distribution of early stages of blue marlin. The distributions of adult blue marlin during their spawning period (May-August) may be used to estimate where larvae might occur (Fig. 28). Under such an assumption, larvae and juveniles would be present throughout all four study zones seaward of the 200 m isobath.

4.4.2 White Marlin (*Tetrapturus albidus*)
4.4.2.1 Adult Distributions

White marlin occur between 35°S and 45°N latitude in the Atlantic, the Caribbean Sea and the Gulf of Mexico (Southeast Fisheries Science Center, 1992). Adults generally occur in water deeper than 100 m when surface temperatures exceed 22 °C (Southeast Fisheries Science Center, 1992). White marlin from the Gulf of Mexico are believed to over-winter off Venezuela and move into the Gulf of Mexico and Atlantic towards feeding grounds as the waters warm. During midsummer, fish are concentrated near the mouth of the Mississippi River followed by a dispersion to other parts of the Gulf in late-summer (Hoese and Moore, 1998).

NMFS longline data reveal the presence of scattered white marlin throughout all study zones from January through April seaward of the 200 m isobath (Fig. 29). The waters in the central region of the eastern zone generally produced few fish. From May through August, white marlin increased in numbers taken and expanded their distribution to encompass most of the study area, although the western third of the western zone and the south-central region of the eastern zone continued to produce the fewest fish (Fig. 29). From September through November, the numbers of fish taken diminished though scattered fish throughout most of the study area. White marlin appeared to move to the south away from the 200 m isobath. By December, the most fish were in the southern half of the western zone and the southwestern part of the central zone.

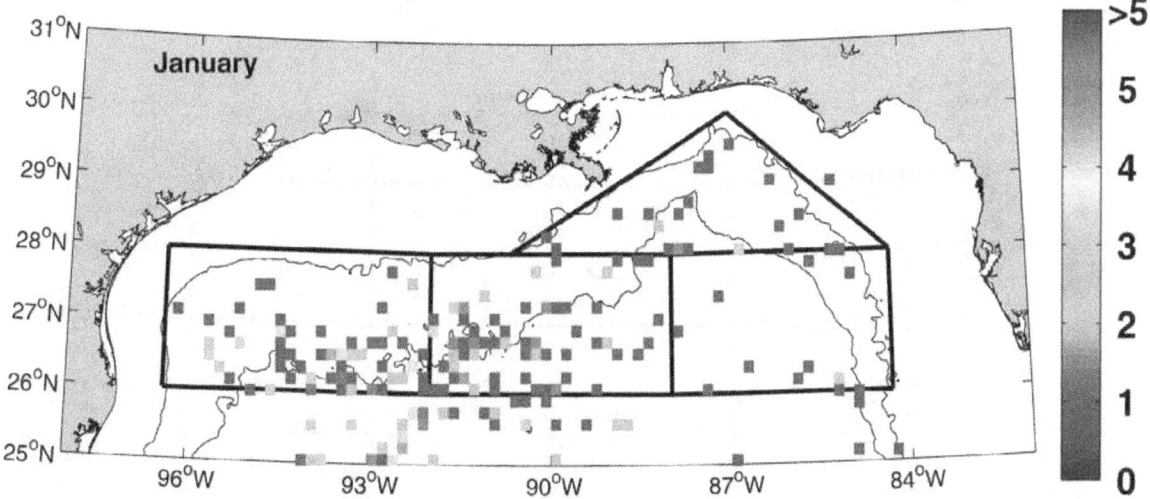

Figure 29. Catch per unit effort of adult white marlin from the commercial long-line fishery. Each square represents the mean catch-per-unit-effort (white marlin per set) taken within a 10' x10' region from January through December over the period 1986-1999. Note that the colorbar was arbitrarily limited to a maximum CPUE of five. The maximum CPUEs were: January = 5.5; February = 5.0; March = 20; April = 11; May = 13; June = 35; July = 45; August = 57; September = 21; October = 21; November =7, December =11.

79

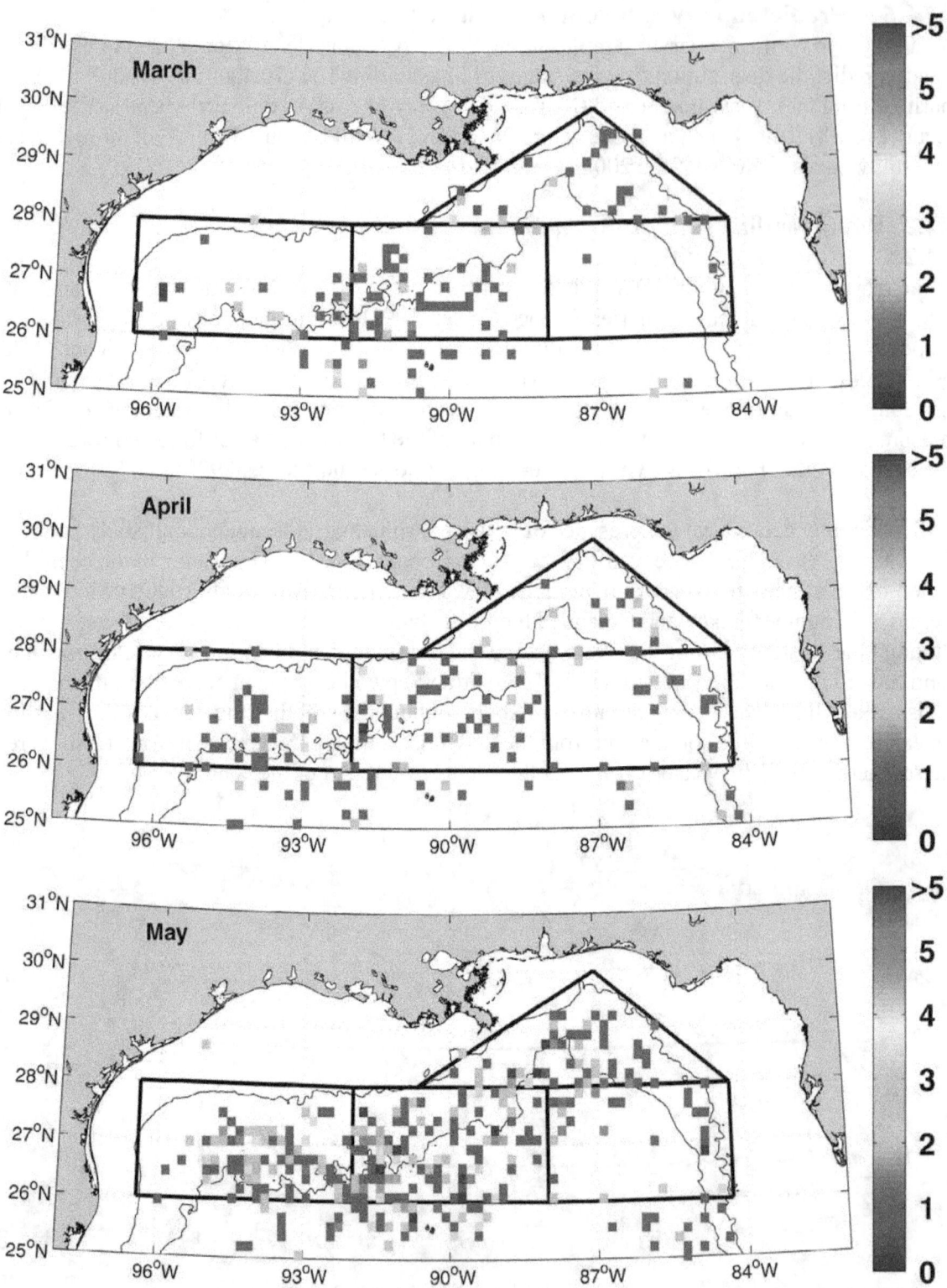

Figure 29. Catch per unit effort of adult white marlin from the commercial long-line fishery. Each square represents the mean catch-per-unit-effort (white marlin per set) taken within a 10' x10' region from January through December over the period 1986-1999. Note that the colorbar was arbitrarily limited to a maximum CPUE of five. The maximum CPUEs were: January = 5.5; February = 5.0; March = 20; April = 11; May = 13; June = 35; July = 45; August = 57; September = 21; October = 21; November =7, December =11. (continued)

80

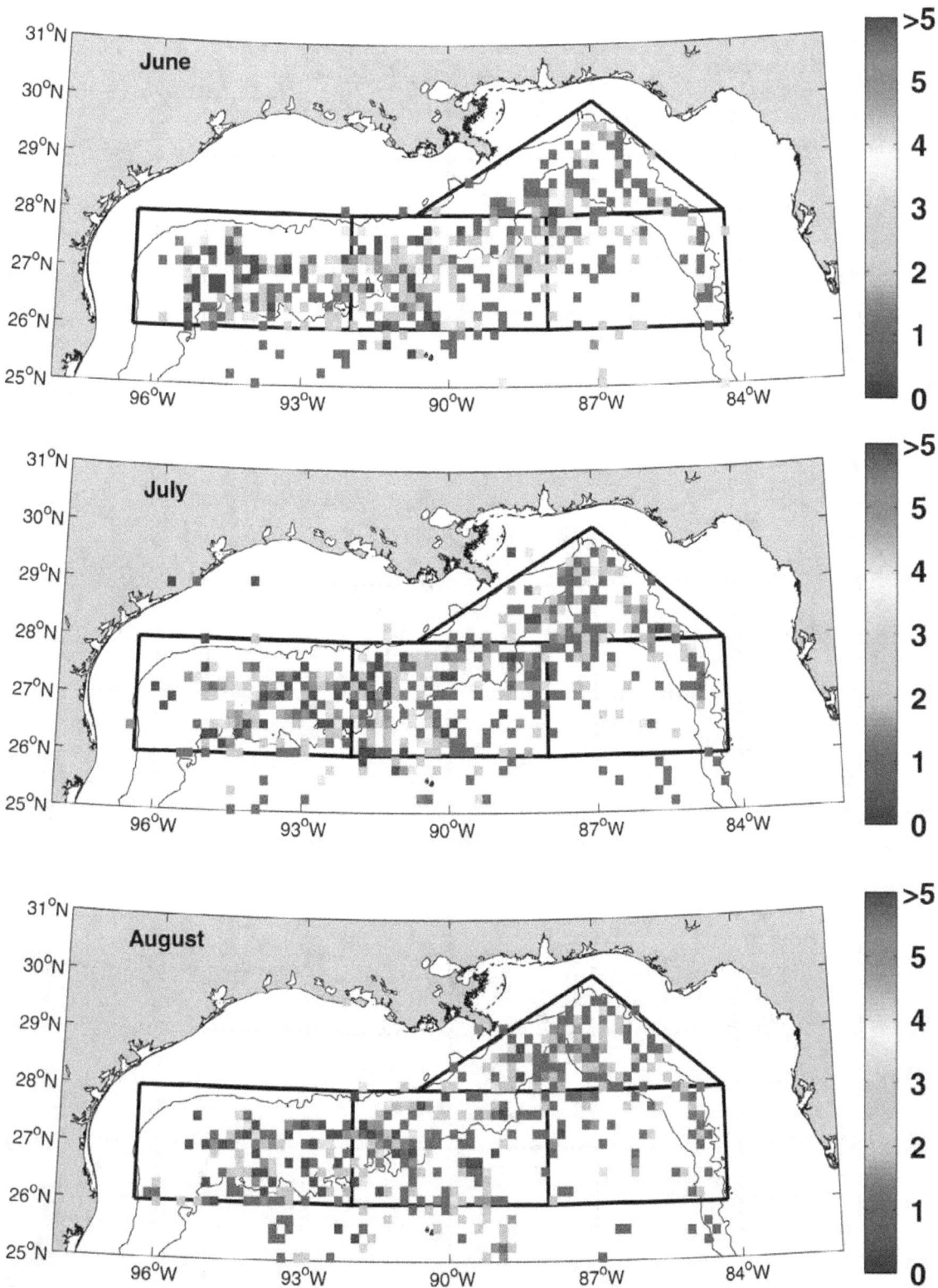

Figure 29. Catch per unit effort of adult white marlin from the commercial long-line fishery. Each square represents the mean catch-per-unit-effort (white marlin per set) taken within a 10' x10' region from January through December over the period 1986-1999. Note that the colorbar was arbitrarily limited to a maximum CPUE of five. The maximum CPUEs were: January = 5.5; February = 5.0; March = 20; April = 11; May = 13; June = 35; July = 45; August = 57; September = 21; October = 21; November =7, December =11. (continued)

81

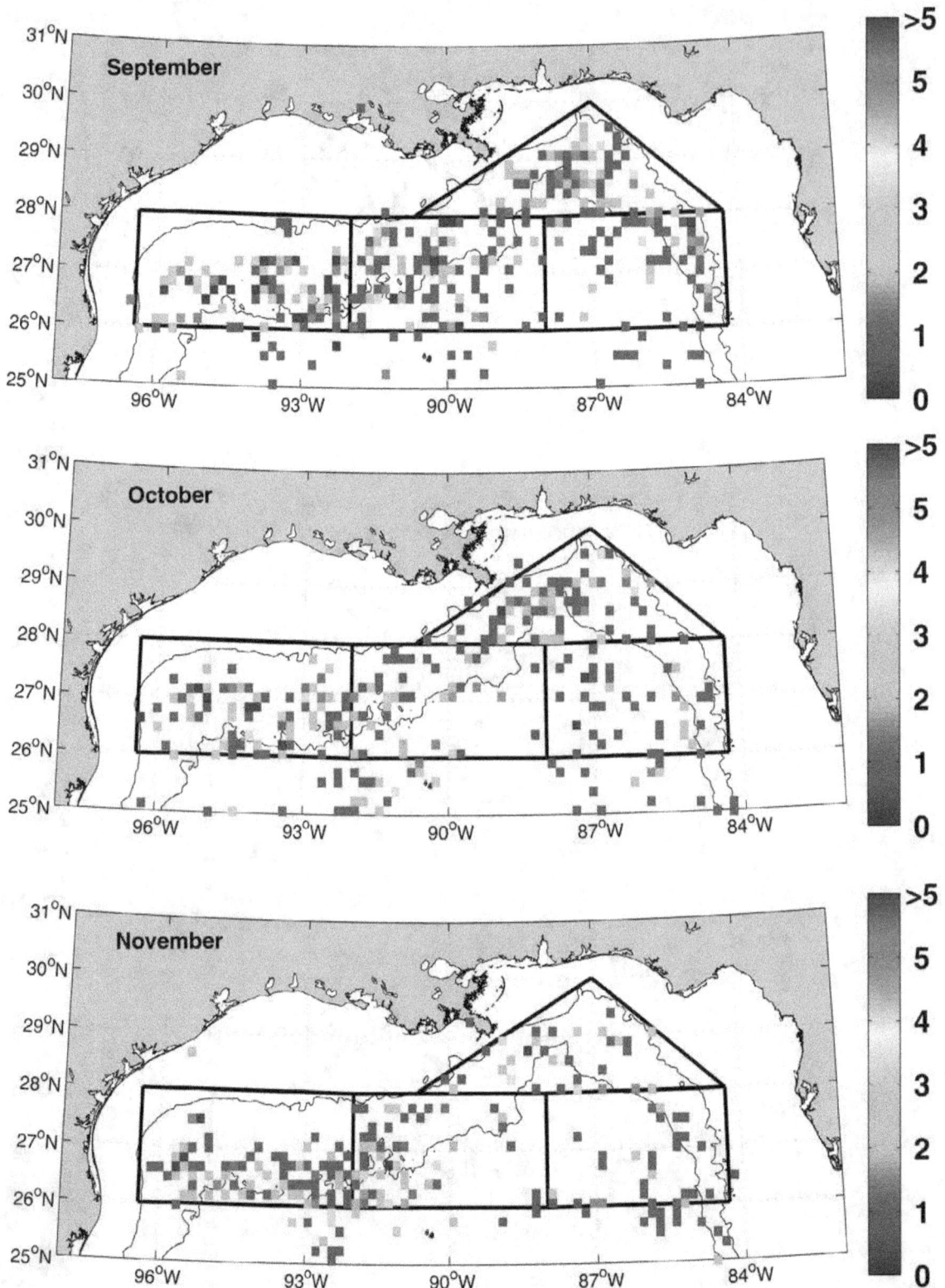

Figure 29. Catch per unit effort of adult white marlin from the commercial long-line fishery. Each square represents the mean catch-per-unit-effort (white marlin per set) taken within a 10' x10' region from January through December over the period 1986-1999. Note that the colorbar was arbitrarily limited to a maximum CPUE of five. The maximum CPUEs were: January = 5.5; February = 5.0; March = 20; April = 11; May = 13; June = 35; July = 45; August = 57; September = 21; October = 21; November =7, December =11. (continued)

82

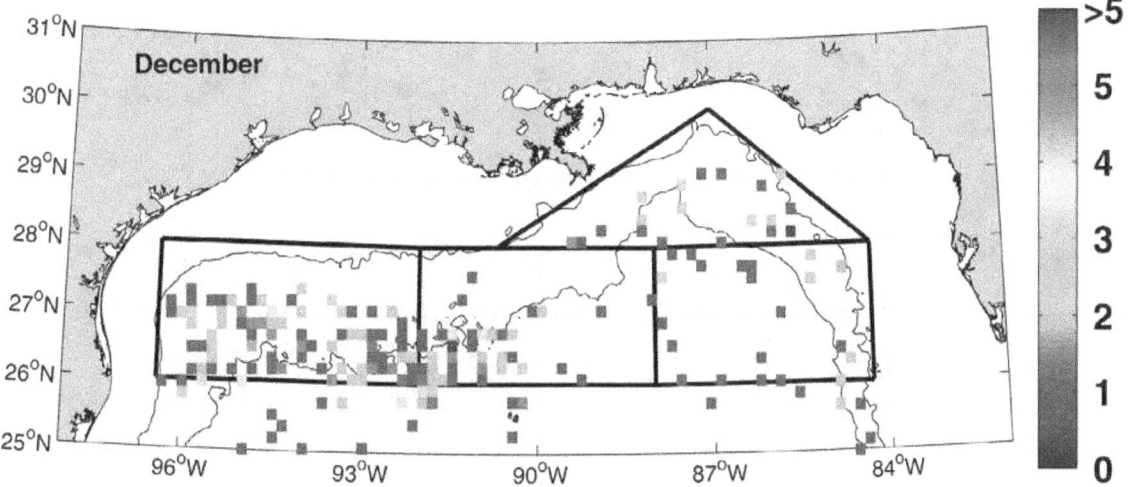

Figure 29. Catch per unit effort of adult white marlin from the commercial long-line fishery. Each square represents the mean catch-per-unit-effort (white marlin per set) taken within a 10' x10' region from January through December over the period 1986-1999. Note that the colorbar was arbitrarily limited to a maximum CPUE of five. The maximum CPUEs were: January = 5.5; February = 5.0; March = 20; April = 11; May = 13; June = 35; July = 45; August = 57; September = 21; October = 21; November =7, December =11. (continued)

4.4.2.2 Reproduction

Spawning occurs throughout the Caribbean, in the Gulf of Mexico and the Florida Straits during April and May, although spawning may extend from November through June between 10°-20°N latitude in the Atlantic (Southeast Fisheries Science Center, 1992). Spawning in the northern Gulf is reported to occur during warm months and through the summer in the more tropical portions of their range (NOS, 1985).

4.4.2.3 Larval/Juvenile Distributions

No larval or juvenile white marlin were reported in the SEAMAP database and the literature did not contain any published records of white marlin larvae or juveniles in the study area.

4.4.2.4 Predicted Adult Distributions

Adult white marlin are present throughout most of the western, central and northern zones of the study area during January and February. Gaps in their distribution are present during January and February within the eastern zone (Fig. 30). By March their distribution covers most of the central and northern zones and the central part of the western zone while they are largely absent from all but the northern edge of the eastern zone (Fig. 30). In April they cover most of all four zones with the exception of a large gap located on the western side of the eastern zone. This gap contracts during May and by June, their distribution encompasses all four zones (Fig. 30). This pattern of complete coverage continues through September. In October they still cover most of all four zones with the exception of a gap in the southeastern part of the central zone. During November and December, they are found throughout most of the study area, however, they shift south of the 200 m isobath and several gaps appear in the central and eastern zones (Fig. 30).

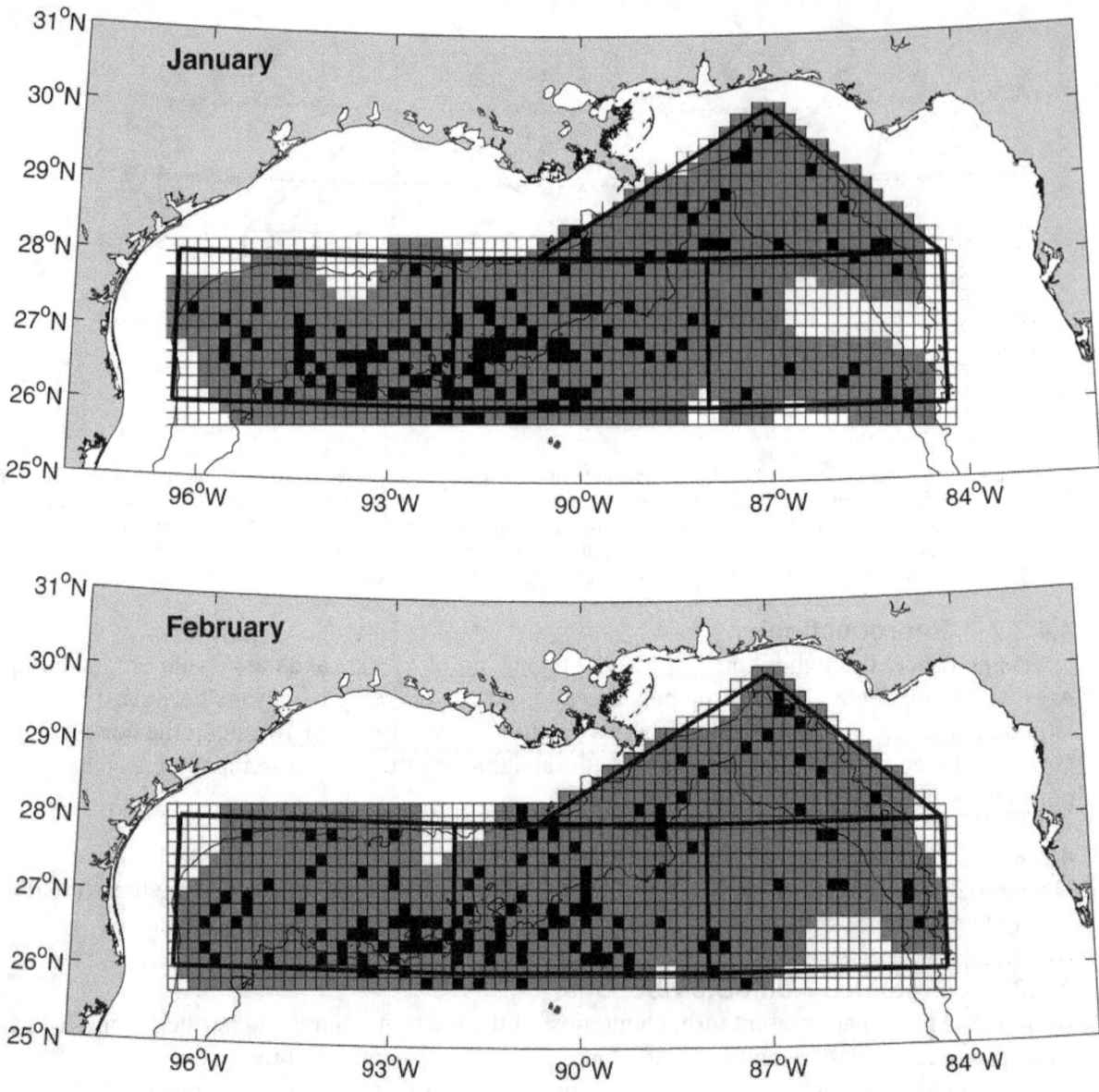

Figure 30. Predicted distributions of adult white marlin in the study area from January through December. The presence of individuals in each grid cell is coded as: confirmed (■), reasonable inference (▨), or unreported (□).

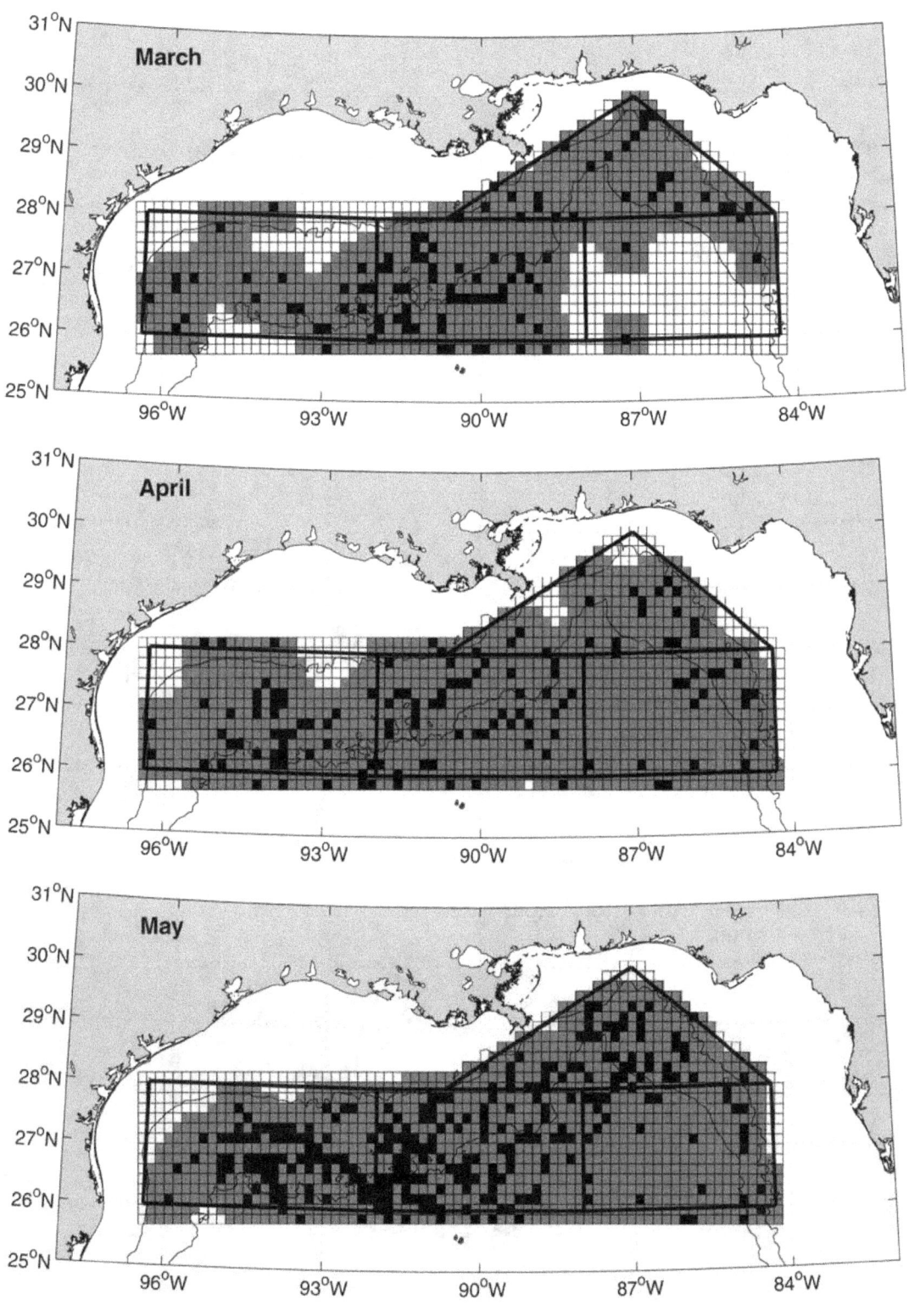

Figure 30.   Predicted distributions of adult white marlin in the study area from January through December. The presence of individuals in each grid cell is coded as: confirmed (■), reasonable inference (▨), or unreported (□). (continued)

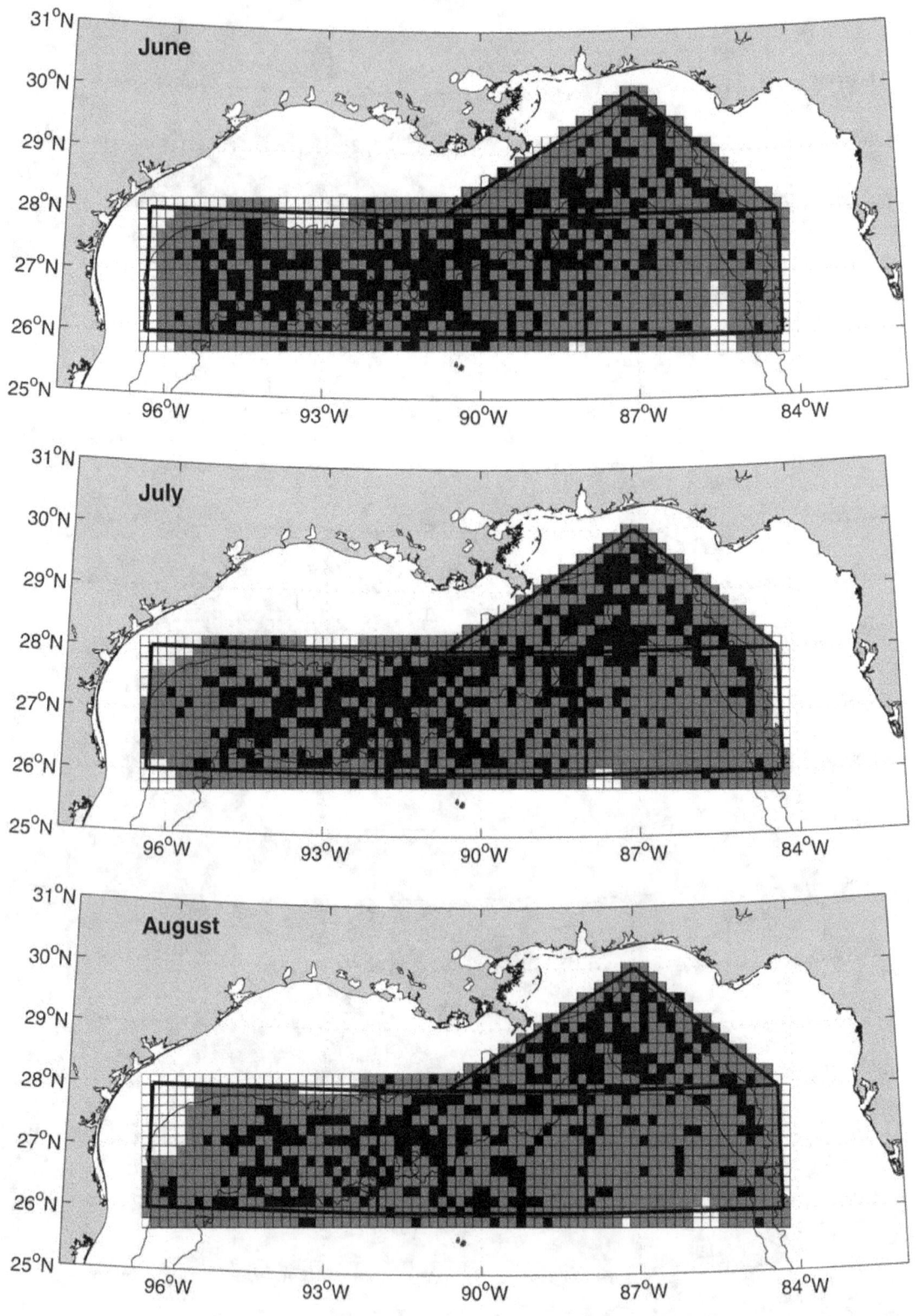

Figure 30. Predicted distributions of adult white marlin in the study area from January through December. The presence of individuals in each grid cell is coded as: confirmed (■), reasonable inference (▒), or unreported (□). (continued)

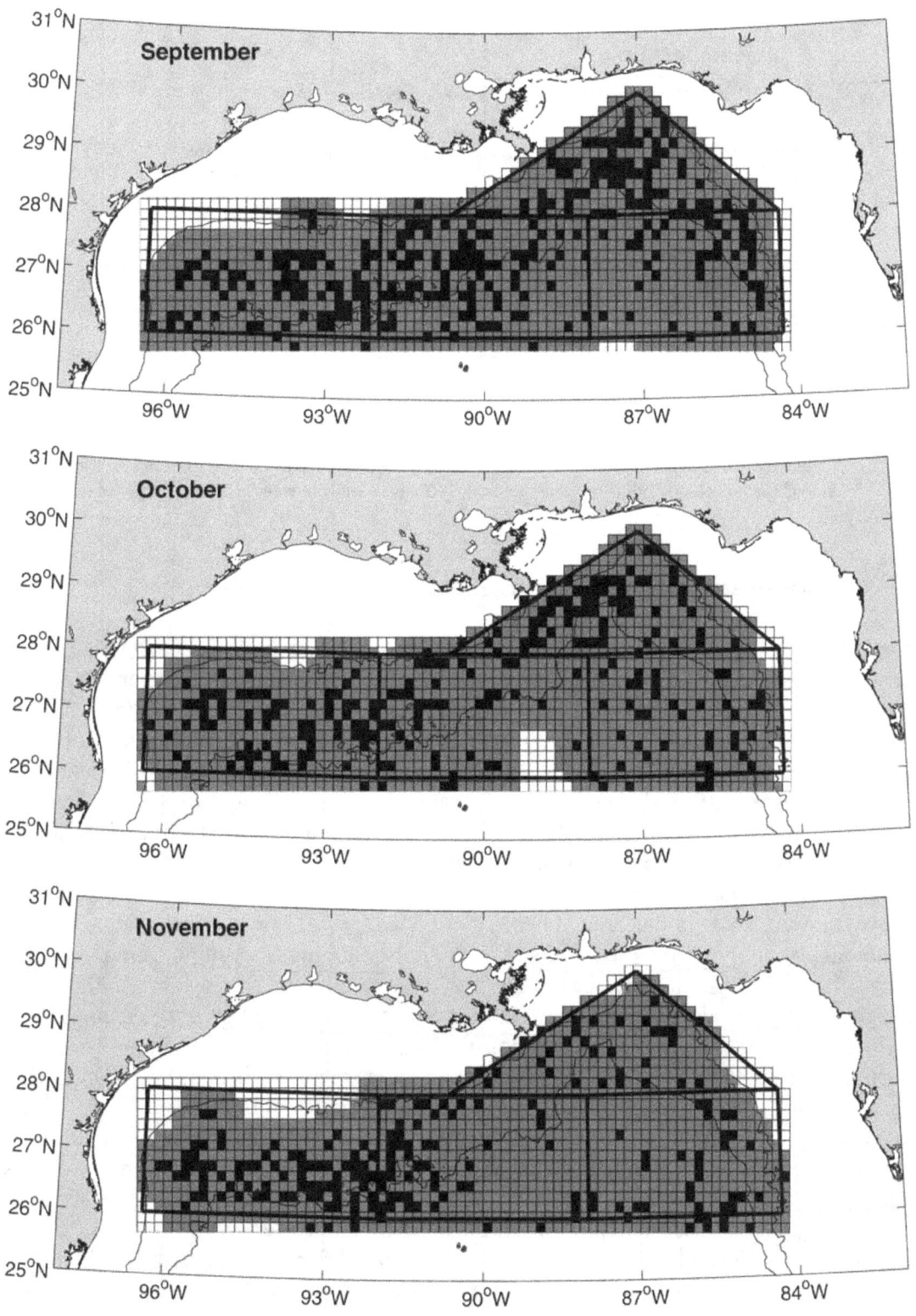

Figure 30. Predicted distributions of adult white marlin in the study area from January through December. The presence of individuals in each grid cell is coded as: confirmed (■), reasonable inference (▨), or unreported (□). (continued)

87

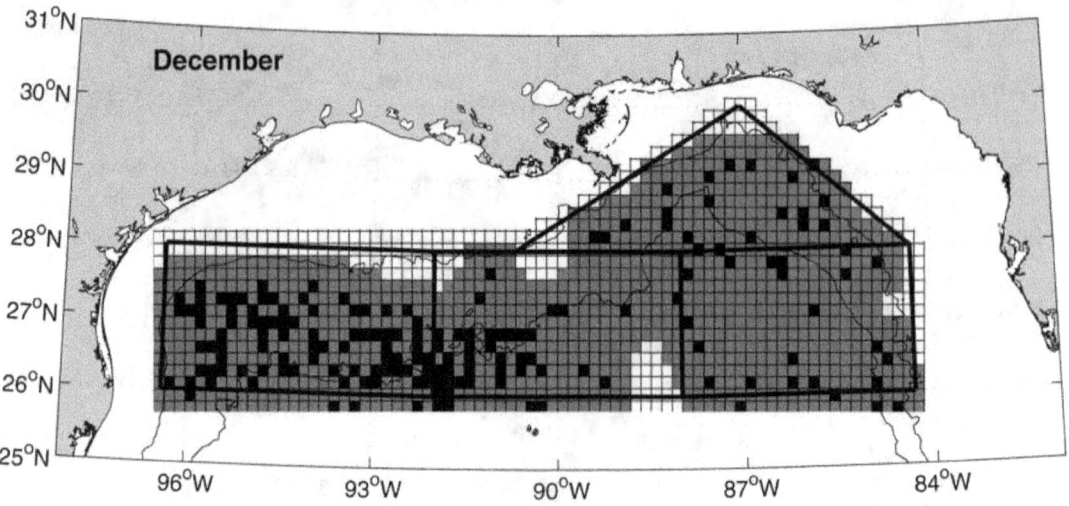

Figure 30. Predicted distributions of adult white marlin in the study area from January through December. The presence of individuals in each grid cell is coded as: confirmed (■), reasonable inference (▨), or unreported (□). (continued)

4.5 Wahoo (*Acanthocybium solanderi*)
4.5.1 Adult Distributions
Wahoo are an offshore pelagic species with a circumtropical distribution occurring in the western Atlantic south from New Jersey to the Caribbean off Columbia (Hoese and Moore, 1998). Information on the distribution of wahoo in the Gulf of Mexico is sparse though the species occurs off northwest Florida, Louisiana, and both north and south Texas (Brusher and Palko, 1987; Cramer, 1995). Franks et al. (2001) report that wahoo are present throughout the year in the northern Gulf of Mexico but are most abundant from spring through fall. The NMFS longline database confirms the broad distribution of wahoo in the northern Gulf of Mexico. Wahoo were present throughout the western, central and northern zones from January through March (Fig. 31). Although there were scattered landings in the eastern zone during December, very few fish were taken in that area in February or March (Fig. 31). In April, wahoo were primarily taken over the waters seaward of the 200 m isobath in the western, central and the western half of the northern zone. Again, few fish were captured in the eastern zone. By May, wahoo were more abundant in the same areas and appeared to expand their distribution throughout the northern zone and into the northwestern corner of the eastern zone. This trend of increasing landings and an expansion into the eastern zone persisted in June (Fig. 31). During July and August wahoo were abundant in all zones except for the south-central region of the eastern zone. Most fish were from waters deeper than 200 m. This distributional pattern continued in September. In October the landings of fish diminished although wahoo were taken throughout most of the study area with lower landings in the northern part of the western zone and the slope water in the eastern portions of the northern and eastern zones (Fig. 31). Landings diminished during November and December when scattered fish were taken in the slope waters of the western, central, and northern zones as well as in scattered locations in the eastern zone (Fig. 31).

88

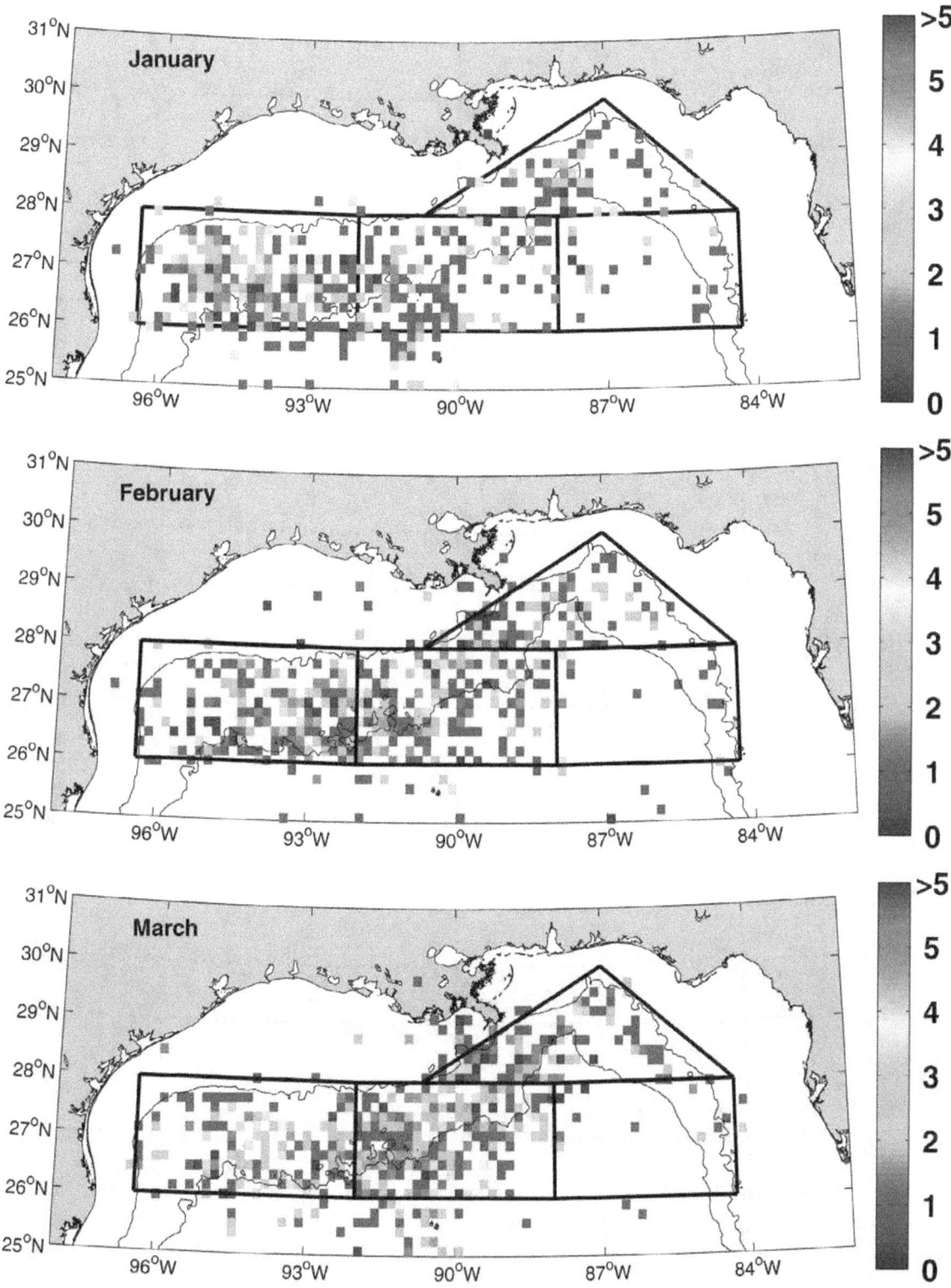

Figure 31. Catch per unit effort of adult wahoo from the commercial long-line fishery. Each square represents the mean catch-per-unit-effort (wahoo per set) taken within a 10' x 10' region from January through December over the period 1986-1999. Note that the colorbar was arbitrarily limited to a maximum CPUE of 5. The maximum CPUEs were: January=47; February = 33.5; March = 56; April = 63; May = 71; June = 37.6; July = 80; August = 100; September = 113; October = 13; November = 21; December = 30.

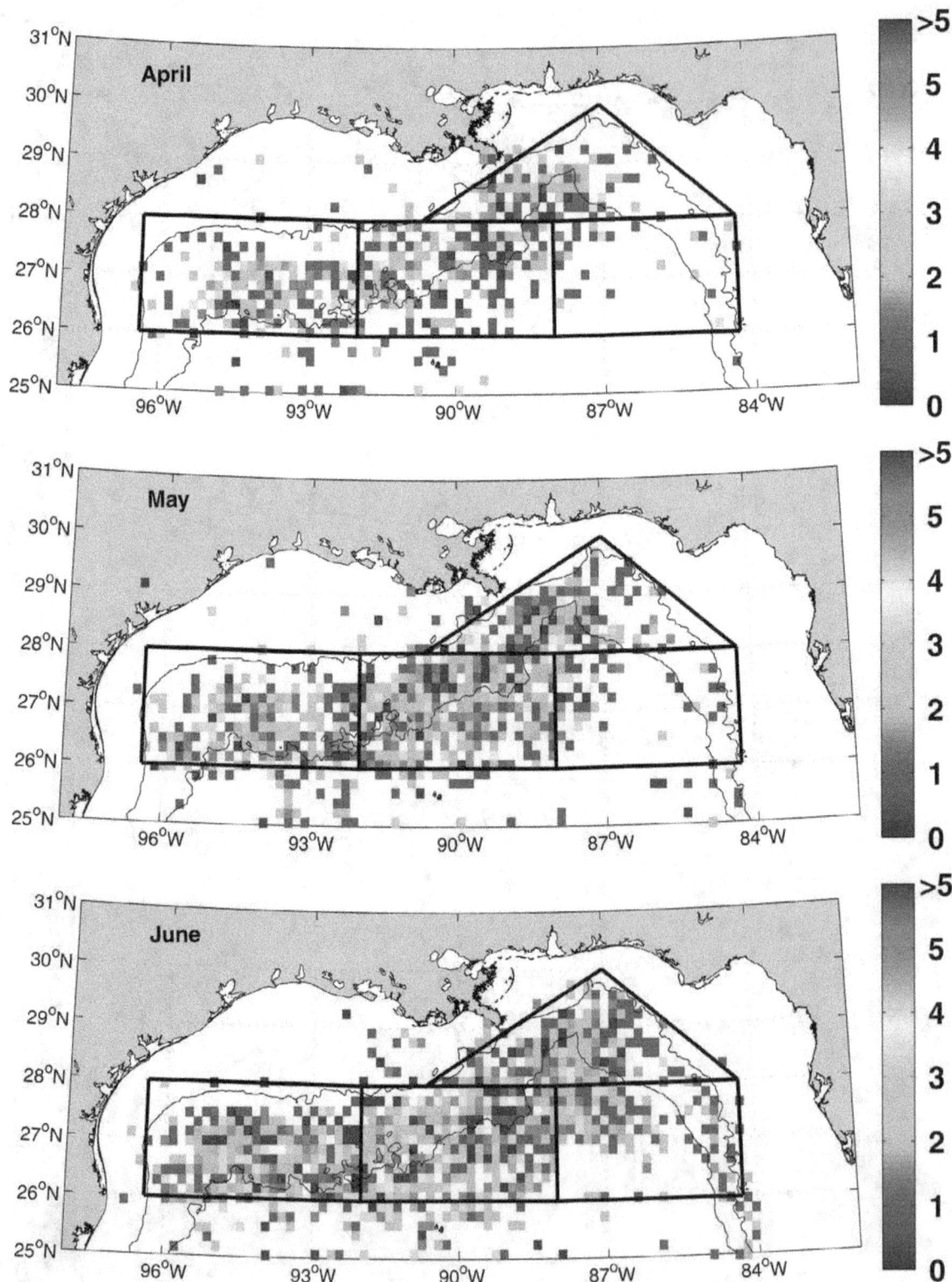

Figure 31. Catch per unit effort of adult wahoo from the commercial long-line fishery. Each square represents the mean catch-per-unit-effort (wahoo per set) taken within a 10' x 10' region from January through December over the period 1986-1999. Note that the colorbar was arbitrarily limited to a maximum CPUE of 5. The maximum CPUEs were: January=47; February = 33.5; March = 56; April = 63; May = 71; June = 37.6; July = 80; August = 100; September = 113; October = 13; November = 21; December = 30. (continued)

90

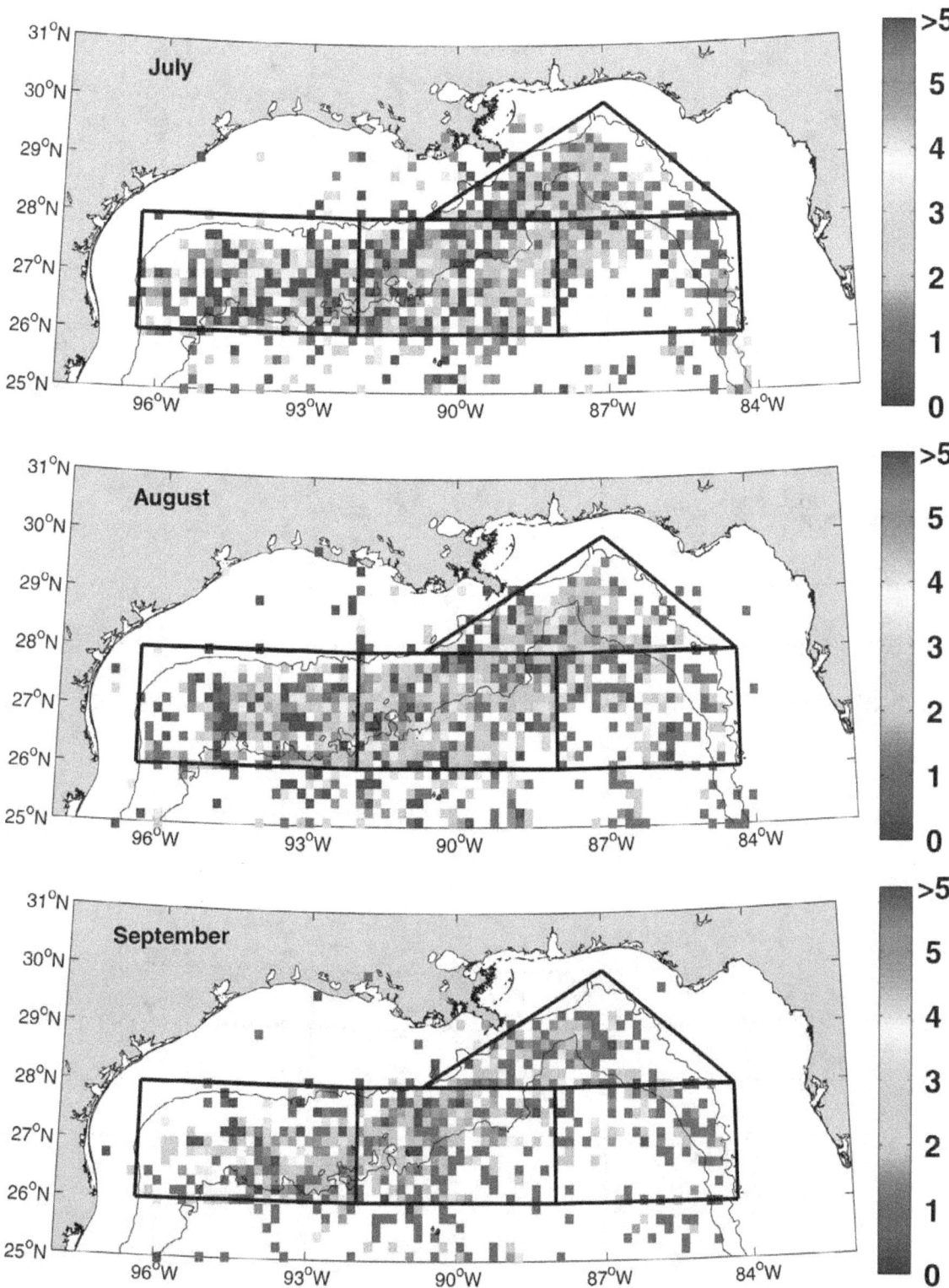

Figure 31. Catch per unit effort of adult wahoo from the commercial long-line fishery. Each square represents the mean catch-per-unit-effort (wahoo per set) taken within a 10' x 10' region from January through December over the period 1986-1999. Note that the colorbar was arbitrarily limited to a maximum CPUE of 5. The maximum CPUEs were: January=47; February = 33.5; March = 56; April = 63; May = 71; June = 37.6; July = 80; August = 100; September = 113; October = 13; November = 21; December = 30. (continued)

91

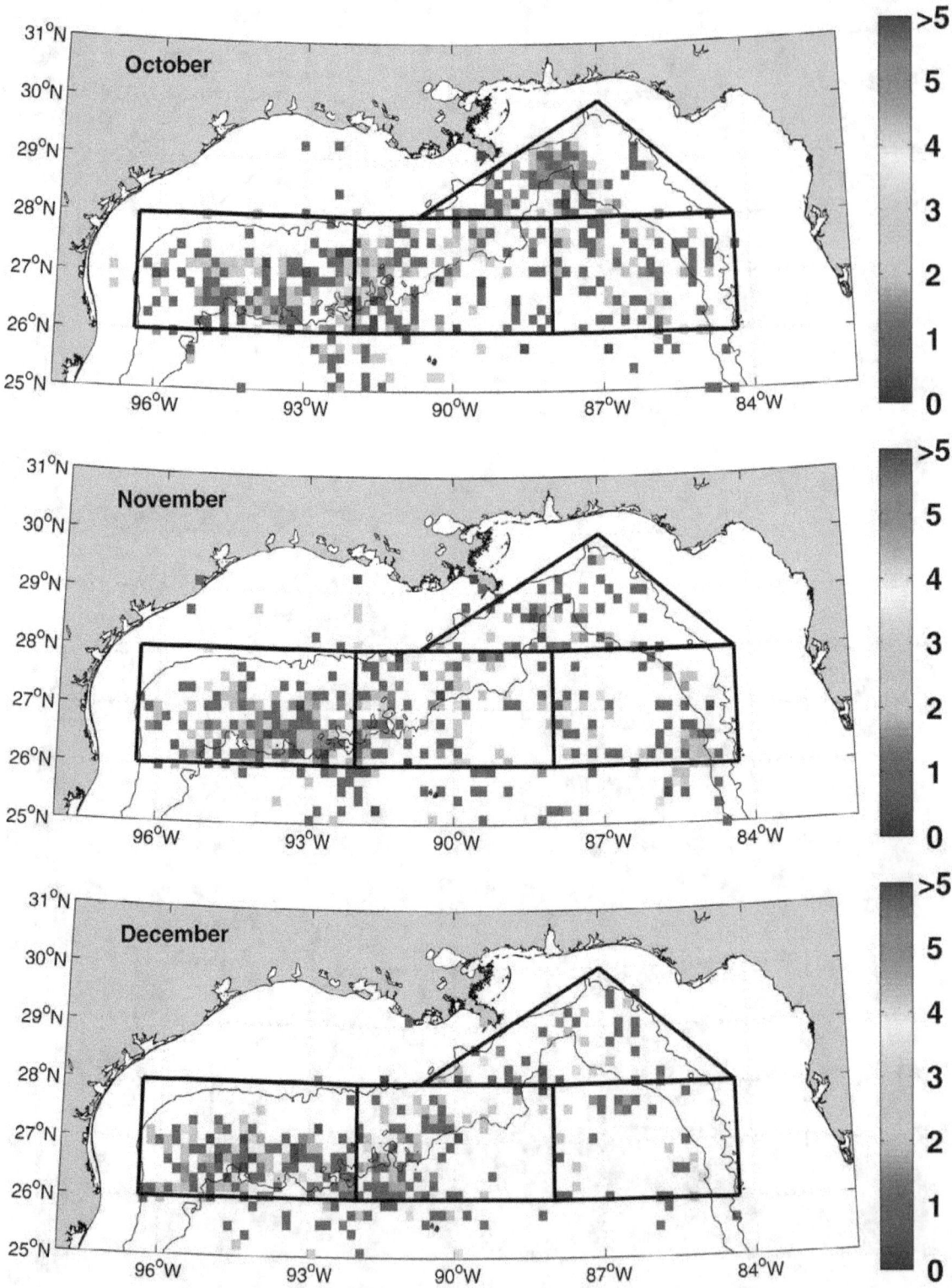

Figure 31. Catch per unit effort of adult wahoo from the commercial long-line fishery. Each square represents the mean catch-per-unit-effort (wahoo per set) taken within a 10' x 10' region from January through December over the period 1986-1999. Note that the colorbar was arbitrarily limited to a maximum CPUE of 5. The maximum CPUEs were: January=47; February = 33.5; March = 56; April = 63; May = 71; June = 37.6; July = 80; August = 100; September = 113; October = 13; November = 21; December = 30. (continued)

4.5.2 Reproduction

Very little information on wahoo reproductive seasonality and localities in the Gulf of Mexico is available in the literature. Recent work by Brown-Peterson et al. (2000) provides the first examination of reproductive seasonality in the Gulf of Mexico. Their results suggest that female wahoo from the northern Gulf of Mexico are capable of spawning from June through August. Fish may be capable of spawning in their second year of life and can produce multiple spawns per season. SEAMAP data contained a few records of wahoo larvae and these were collected during May, August and September suggesting that spawning occurs within approximately the same time range. Additionally, anecdotal information from charterboat captains and recreational anglers published on the world wide web suggests that they spawn from June to August in waters off the southeastern U.S and from May through October with a peak during June in the northern Gulf of Mexico.

4.5.3 Larval/Juvenile Distributions

Wahoo larvae were present in SEAMAP samples collected during May from the slope water of the eastern zone (Fig. 32). Additionally, Richards et al. (1993) also reported a larval wahoo from the slope water of the eastern zone. No larvae were present during June or July but they appeared again in August in the slope water of the central parts of the northern and western zones. In September a single larva was collected on the eastern edge of the eastern zone in the slope water (Fig. 32).

Limited samples of larval wahoo (2 individuals) were collected by Hare et al. (2001) at 20-30 m and 30-40 m at night north of St. Croix suggesting that larvae of this species occur in surface waters. Juveniles likely co-occur with larvae.

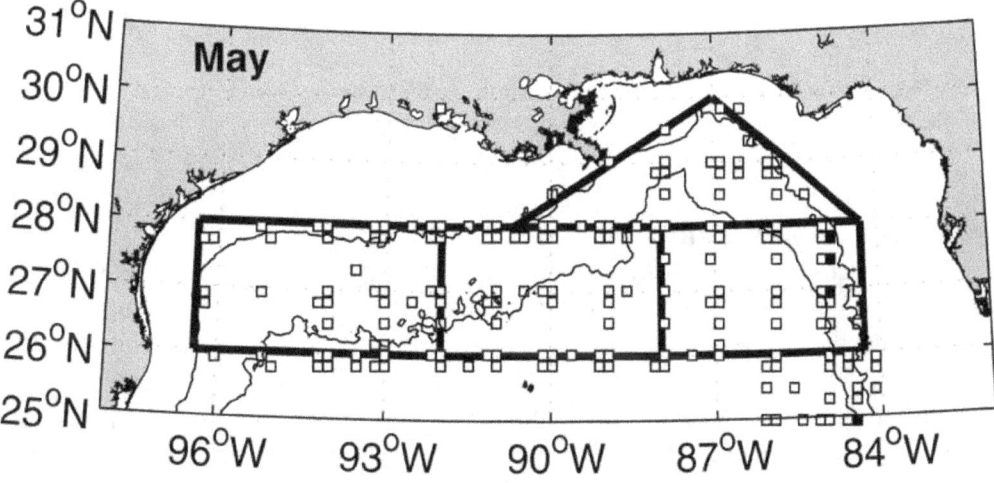

Figure 32.   Presence (■) and absence (□) of wahoo larvae in the study area during May, August, and September. These were the only months during which time, wahoo were present in SEAMAP ichthyoplankton samples.

93

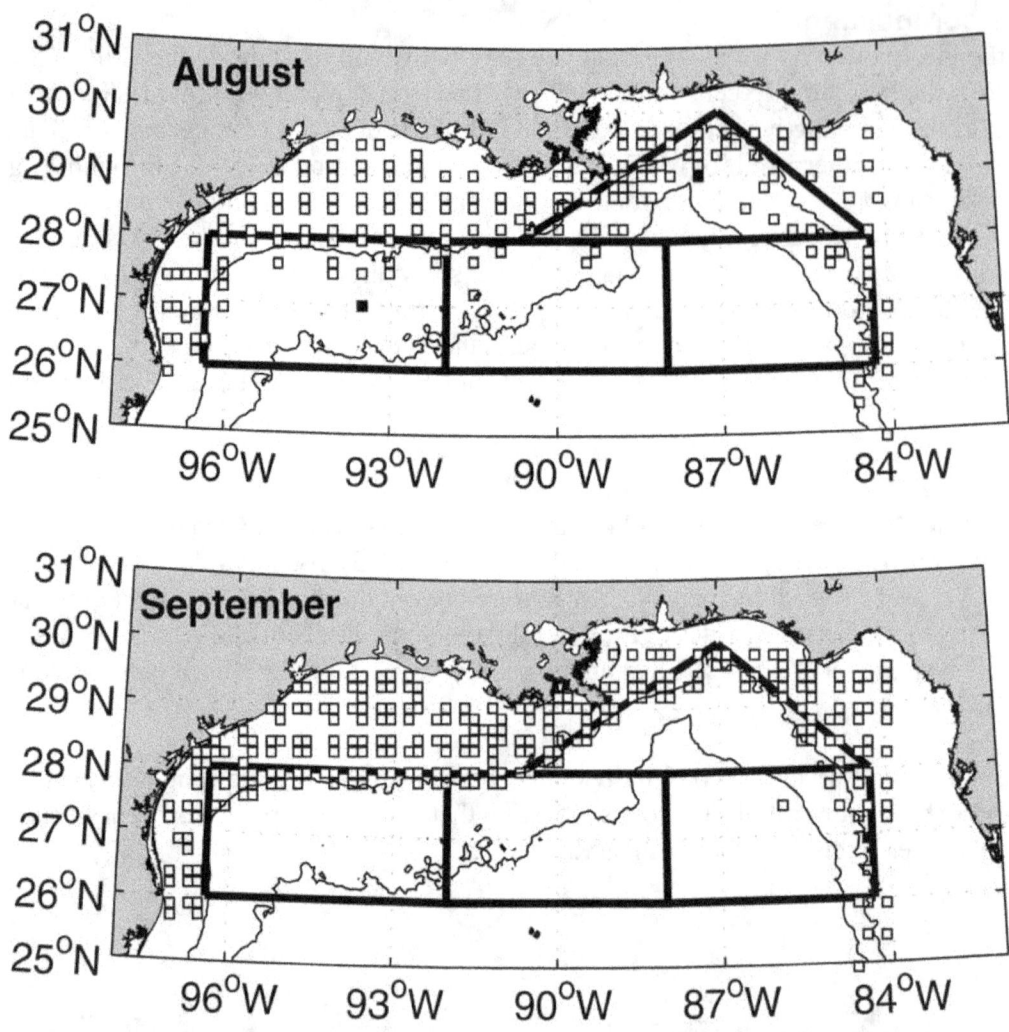

Figure 32. Presence (■) and absence (□) of wahoo larvae in the study area during May, August, and September. These were the only months during which time, wahoo were present in SEAMAP ichthyoplankton samples. (continued)

4.5.4 Predicted Adult Distributions

Adult wahoo are predicted to occur throughout the four study zones during January with the exception of the south-central region of the eastern zone (Fig. 33). During February and March their distribution encompasses all of the northern, western and central zones while the region of absence in the southern portion of the eastern zone expands. The area of the eastern zone where wahoo are absent diminishes through April and May until in June, wahoo occupy the waters of all four zones (Fig. 33). This pattern persists until November when wahoo are no longer present in the waters landward of the 200 m isobath in the northern zone. In December wahoo also move seaward off the shelf in the eastern and western zones and gaps appear in their distribution in the central and eastern zones (Fig. 33).

94

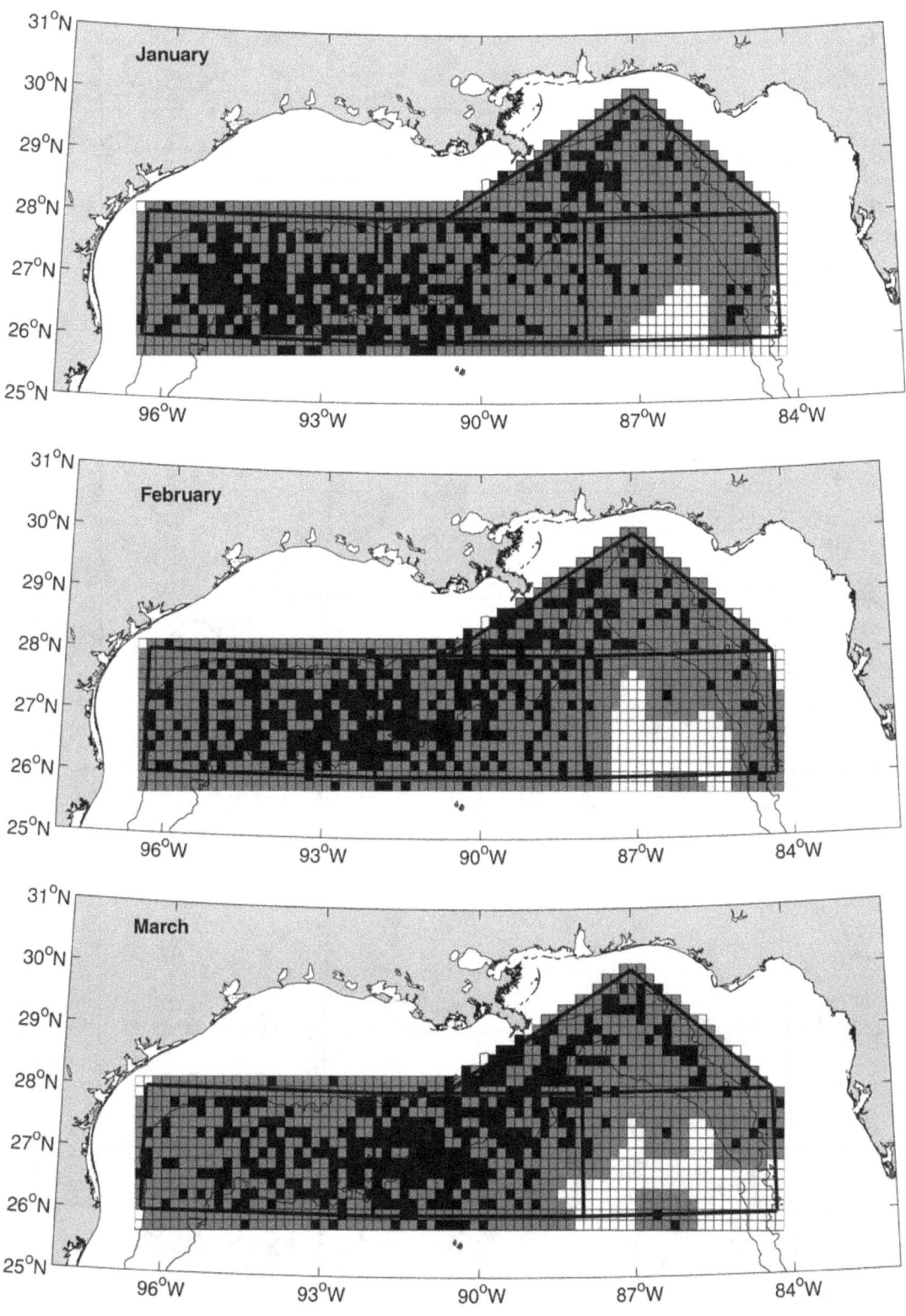

Figure 33. Predicted distributions of adult wahoo in the study area from January through December. The presence of individuals in each grid cell is coded as: confirmed (■), reasonable inference (▨), or unreported (□).

95

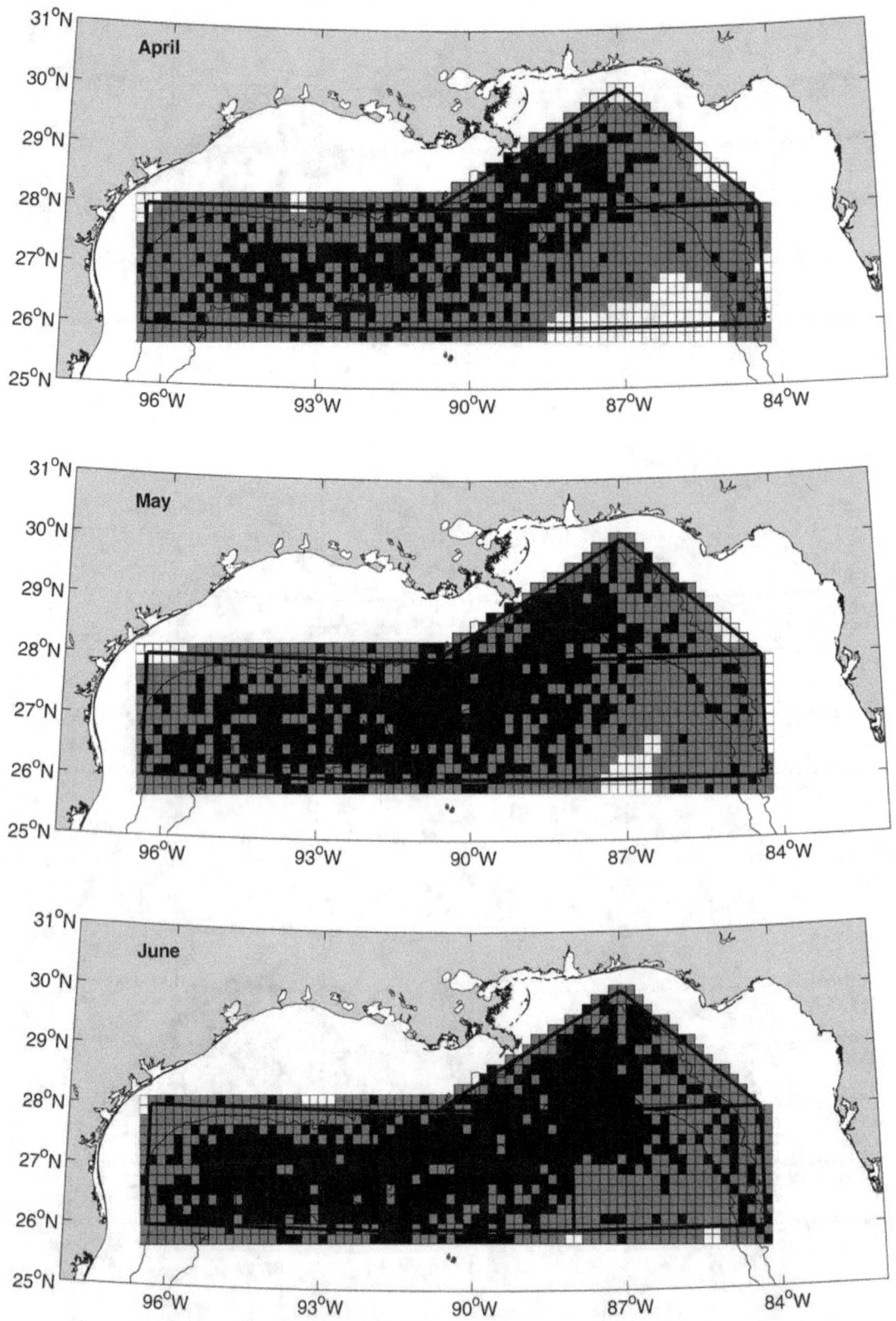

Figure 33. Predicted distributions of adult wahoo in the study area from January through December. The presence of individuals in each grid cell is coded as: confirmed (■), reasonable inference (▨), or unreported (□). (continued)

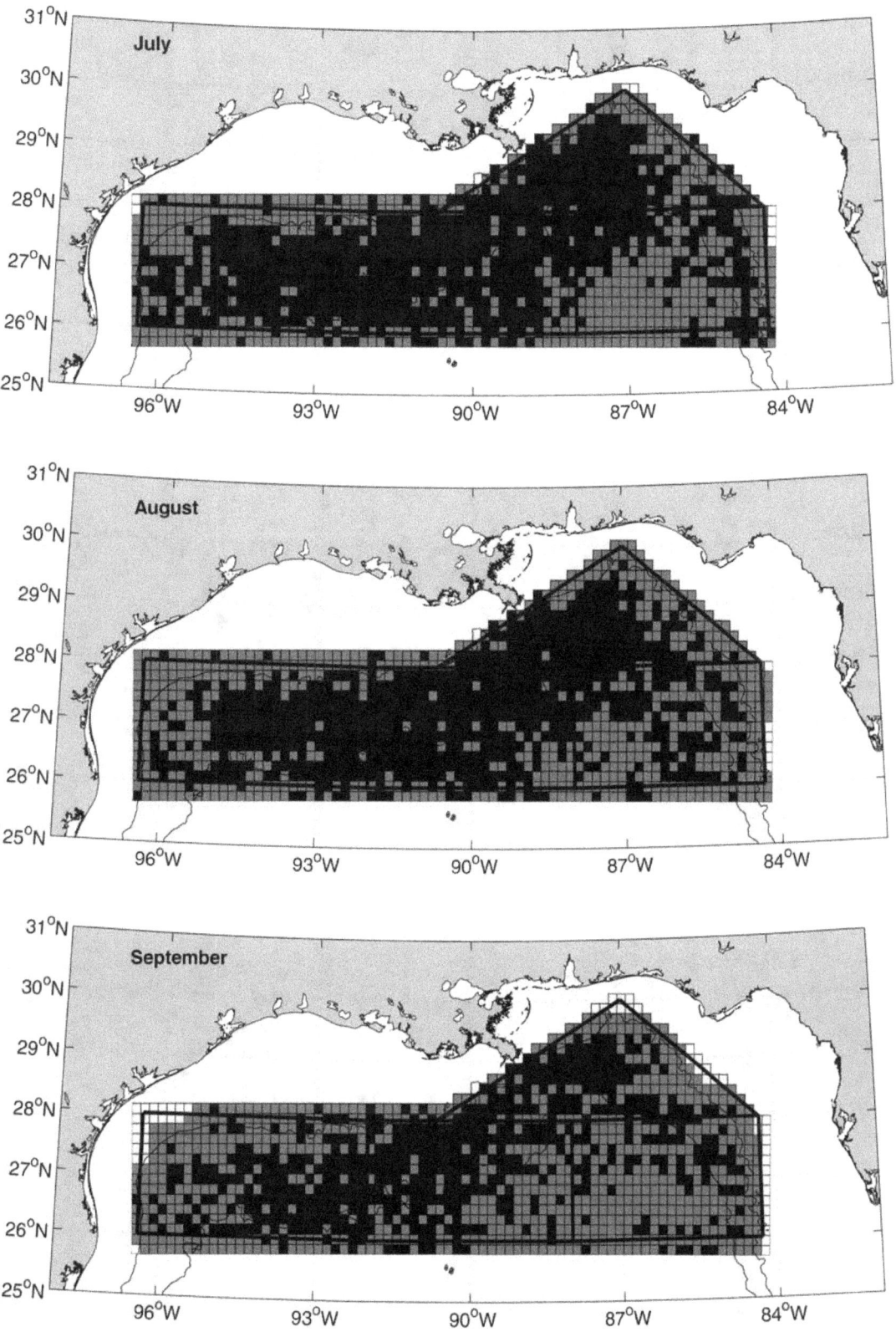

Figure 33. Predicted distributions of adult wahoo in the study area from January through December. The presence of individuals in each grid cell is coded as: confirmed (■), reasonable inference (▨), or unreported (□). (continued)

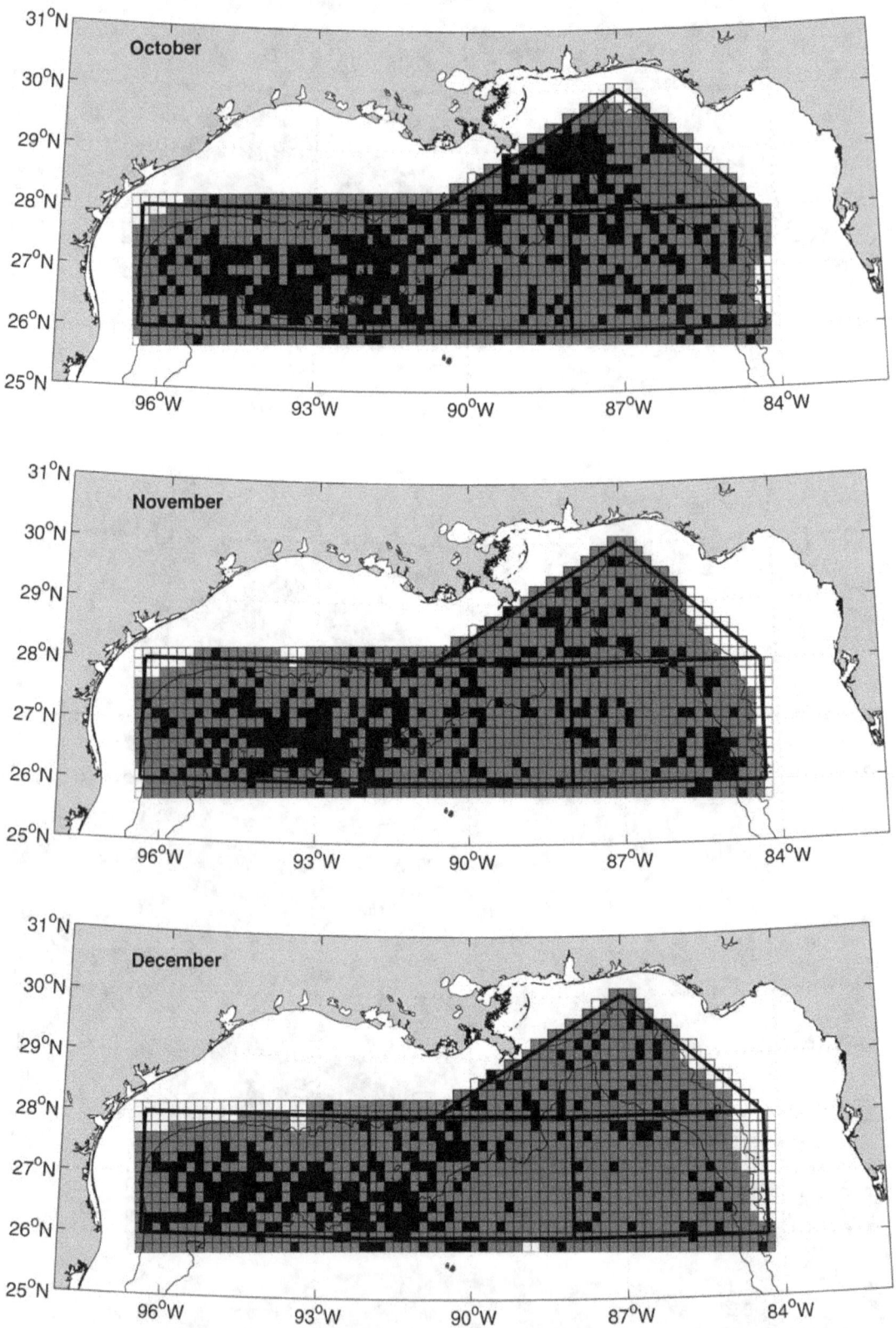

Figure 33. Predicted distributions of adult wahoo in the study area from January through December. The presence of individuals in each grid cell is coded as: confirmed (■), reasonable inference (▨), or unreported (□). (continued)

98

4.5.5   Predicted Larval/Juvenile Distributions

With few records of early stage wahoo available, it is not currently possible to assess their distribution throughout the study area. Since all records were obtained at locations between 200 and 2000 m, (Fig. 34), it appears likely that wahoo larvae and juveniles are at least present throughout the slope water regions of the four study zones.

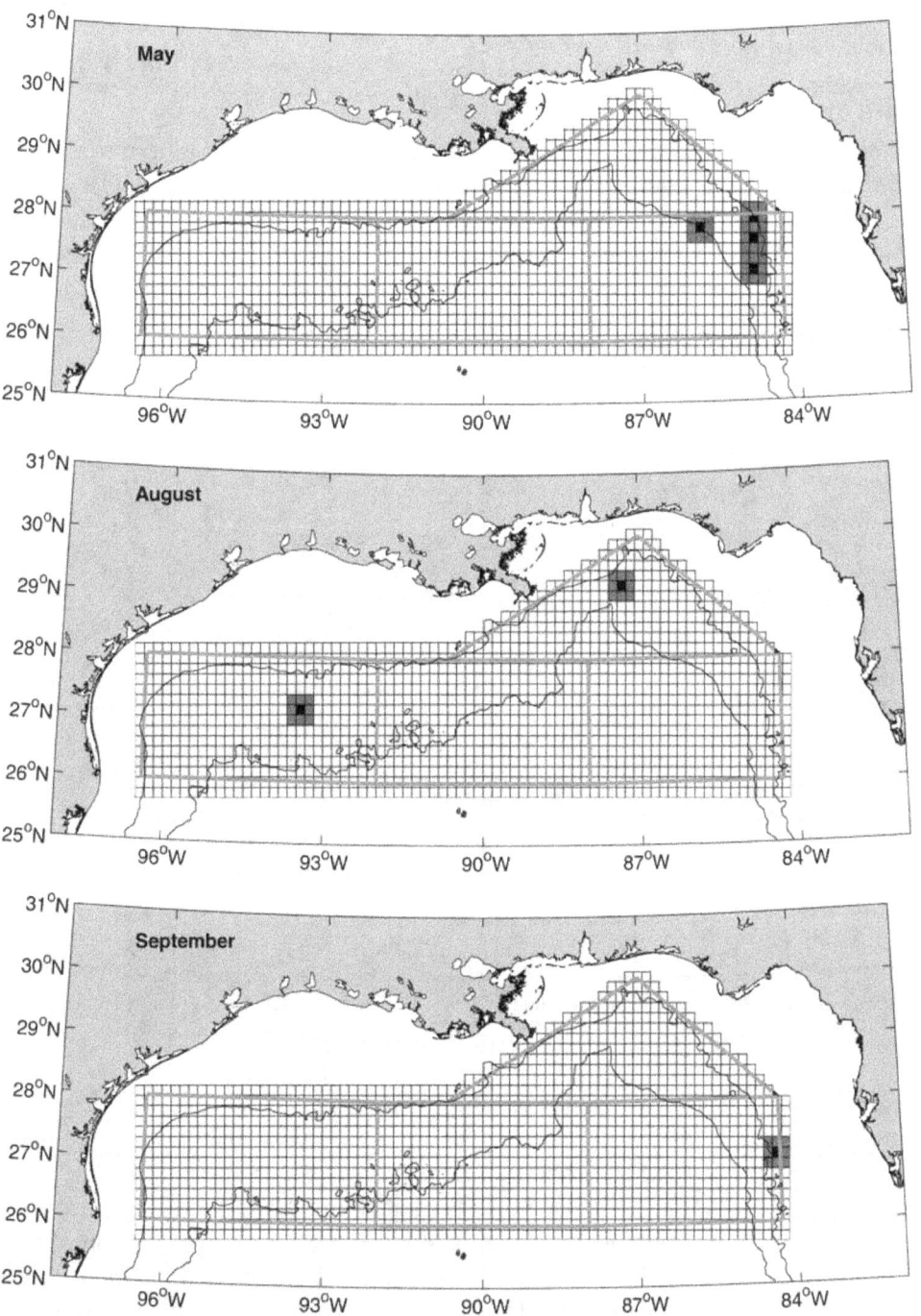

Figure 34.   Predicted distributions of larval/juvenile wahoo in the study area during May, August, and September. An additional observation from a transect conducted by Richards et al. 1993 is included in the May data. The presence of individuals in each grid cell is indicated as confirmed (■), reasonable inference (▨) or unreported (□).

99

4.6 Dolphinfish (*Coryphaena hippurus*)

Dolphinfish are pelagic predators widely distributed throughout the tropical and subtropical oceans. They are frequently associated with flotsam and are also important members of the *Sargassum* community (Beardsley, 1967). Dolphinfish forage around *Sargassum* and *Sargassum* is a common component of the stomach contents of dolphinfish (Manooch et al. 1984). This likely represents material that was incidentally ingested while foraging on fish and invertebrates closely associated with the *Sargassum* community.

4.6.1 Adult Distributions

Coryphaena hippurus is a cosmopolitan species that is distributed throughout tropical and subtropical waters and most adult specimens collected by Gibbs and Collette (1959) were from waters warmer than 21.1 °C (70 °F). This species is common in the Gulf of Mexico and is frequently associated with *Sargassum* rafts (Dooley, 1972; Bortone et al. 1977) and other flotsam (Fisher, 1978). Dolphinfish are reported to move into the coastal waters of the northern Gulf of Mexico during spring and summer months and migrate into more southerly Gulf waters during fall and winter (NOS, 1985) (Fig. 35).

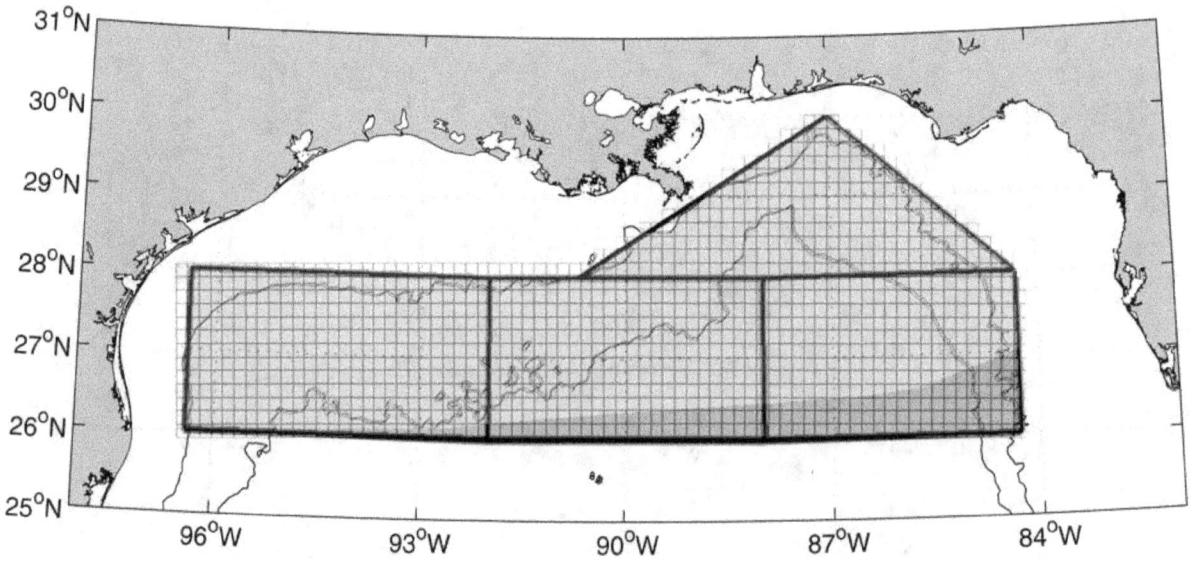

Figure 35.   Seasonal distribution of dophinfish predicted by NOS (1985) for the summer (▪) and winter (▪).

NMFS longline data indicate that catches of dolphinfish during January are generally low and confined to waters seaward of the 200 m isobath (Fig. 36). This distributional pattern did not change much during February and March. Landings increased substantially from April through July during which time dolphinfish move northward and landward of the 200 m isobath. By July they were distributed throughout most of the four study zones with the exception of the oceanic waters in the southern portion of the eastern zone (Fig. 36). Landings diminished from August through December during which time the dolphinfish appear to migrate back into the slope water and oceanic water to the south (Fig. 36).

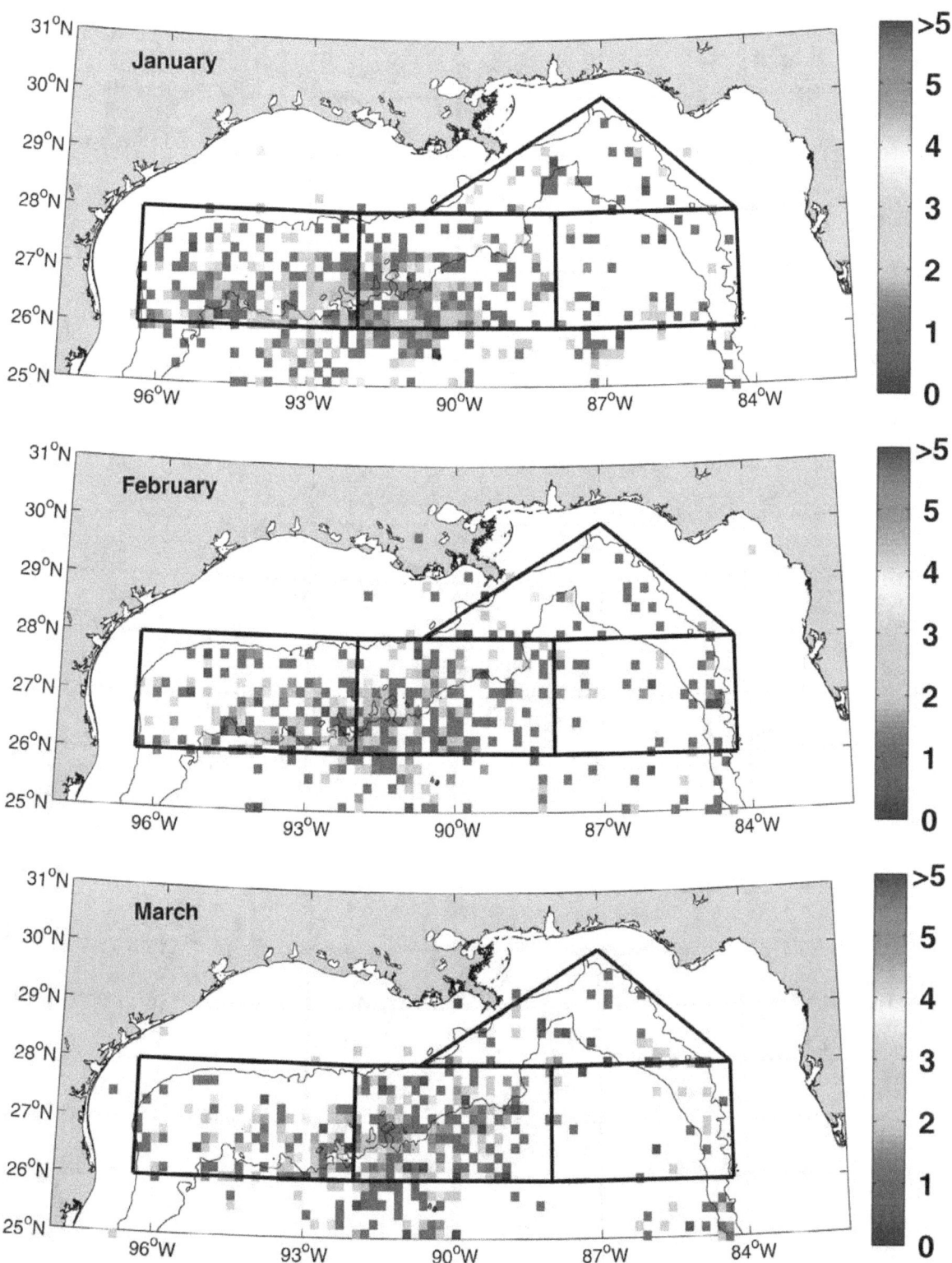

Figure 36. Catch per unit effort of adult dolphinfish from the commercial long-line fishery. Each square represents the mean catch-per-unit-effort (dolphinfish per set) taken within a 10' x10' region from January through December over the period 1986-1999. Note that the colorbar was arbitrarily limited to a maximum CPUE of 5. The maximum CPUEs were: January =27; February =27; March=109; April =70; May =126; June=168.5; July =132.3; August =100; September=132.5; October = 38; November = 67; December=22.

101

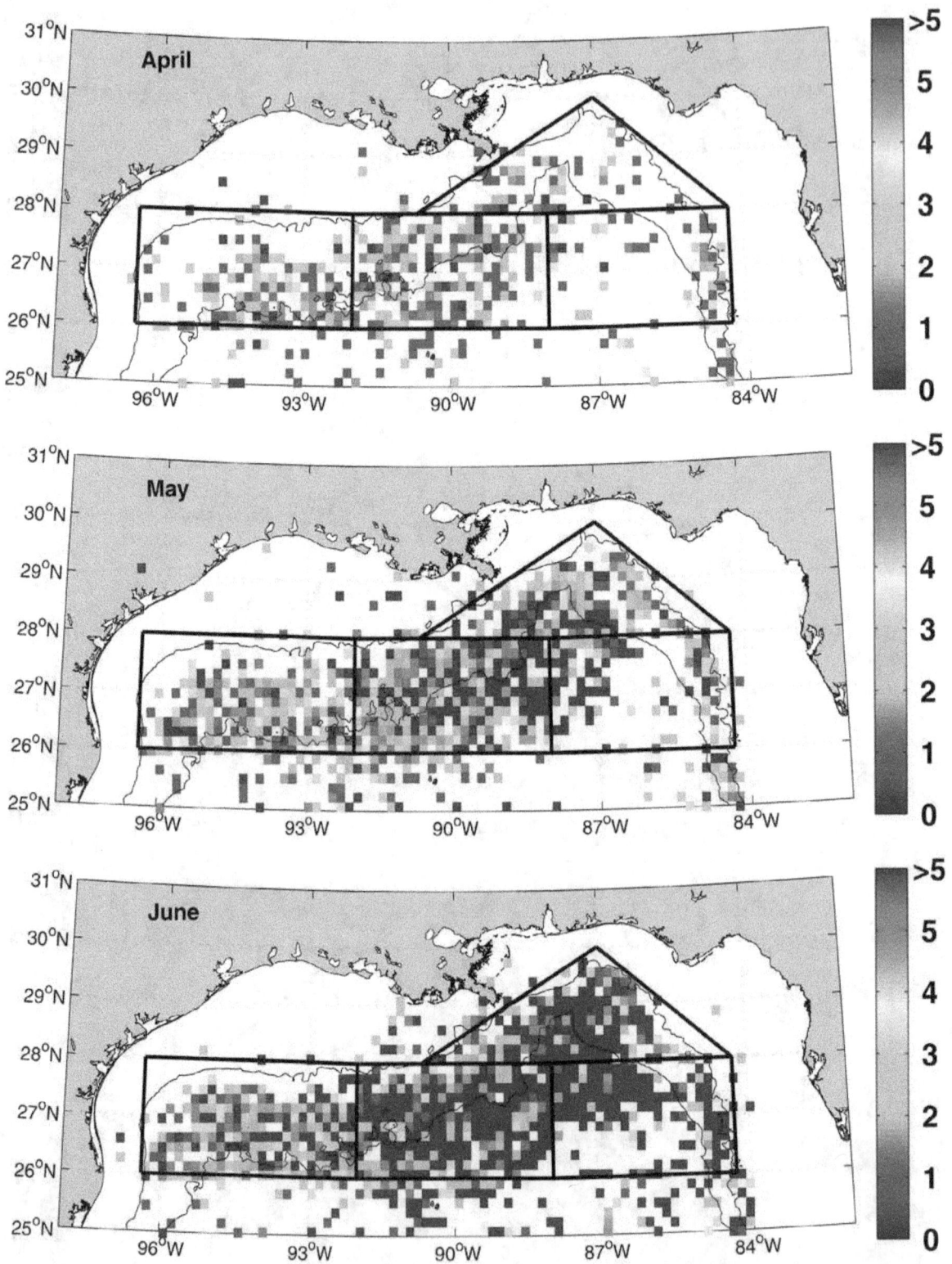

Figure 36. Catch per unit effort of adult dolphinfish from the commercial long-line fishery. Each square represents the mean catch-per-unit-effort (dolphinfish per set) taken within a 10' x10' region from January through December over the period 1986-1999. Note that the colorbar was arbitrarily limited to a maximum CPUE of 5. The maximum CPUEs were: January =27; February =27; March=109; April =70; May =126; June=168.5; July =132.3; August =100; September=132.5; October = 38; November = 67; December=22. (continued)

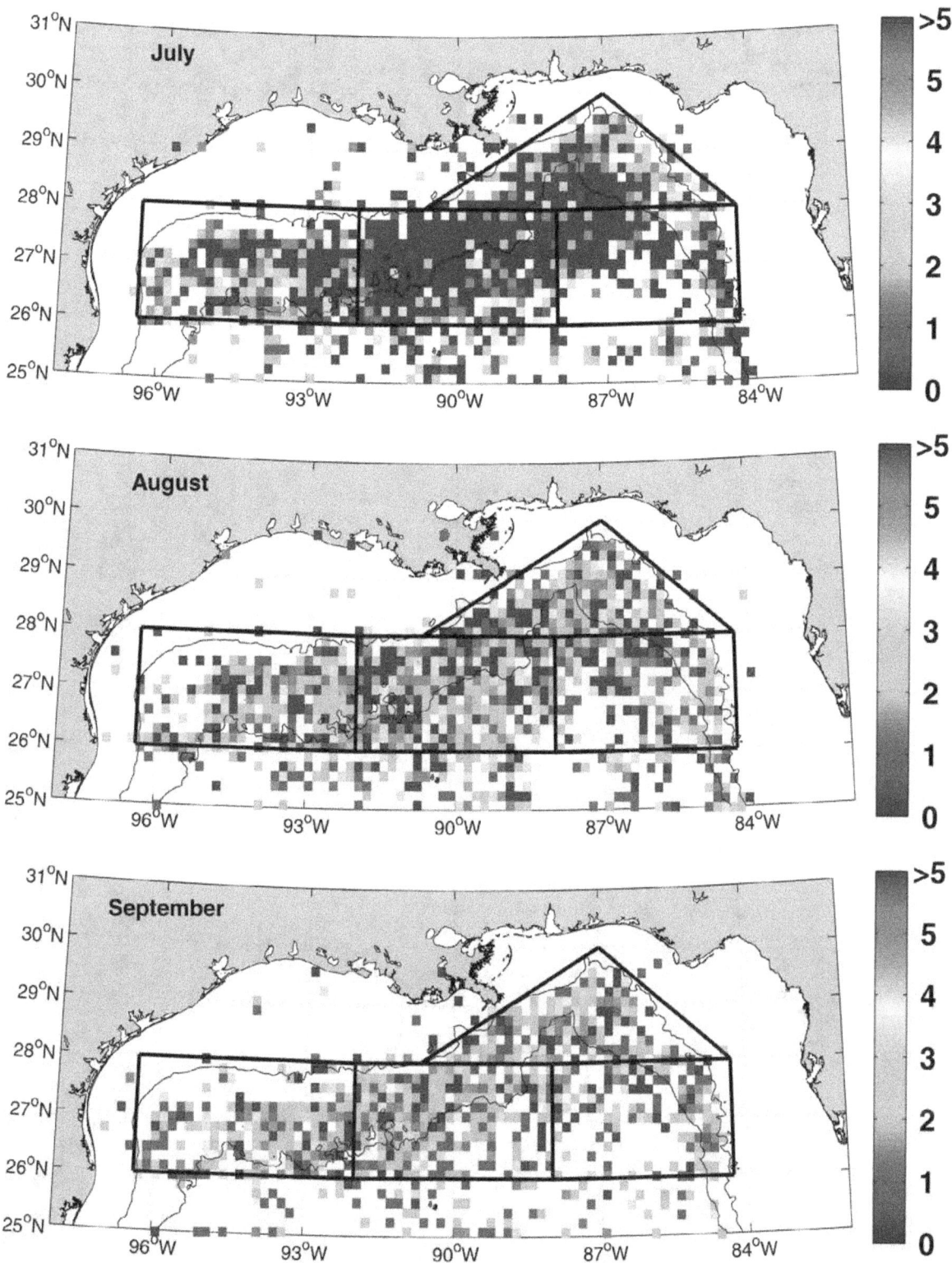

Figure 36. Catch per unit effort of adult dolphinfish from the commercial long-line fishery. Each square represents the mean catch-per-unit-effort (dolphinfish per set) taken within a 10' x10' region from January through December over the period 1986-1999. Note that the colorbar was arbitrarily limited to a maximum CPUE of 5. The maximum CPUEs were: January =27; February =27; March=109; April =70; May =126; June=168.5; July =132.3; August =100; September=132.5; October = 38; November = 67; December=22. (continued)

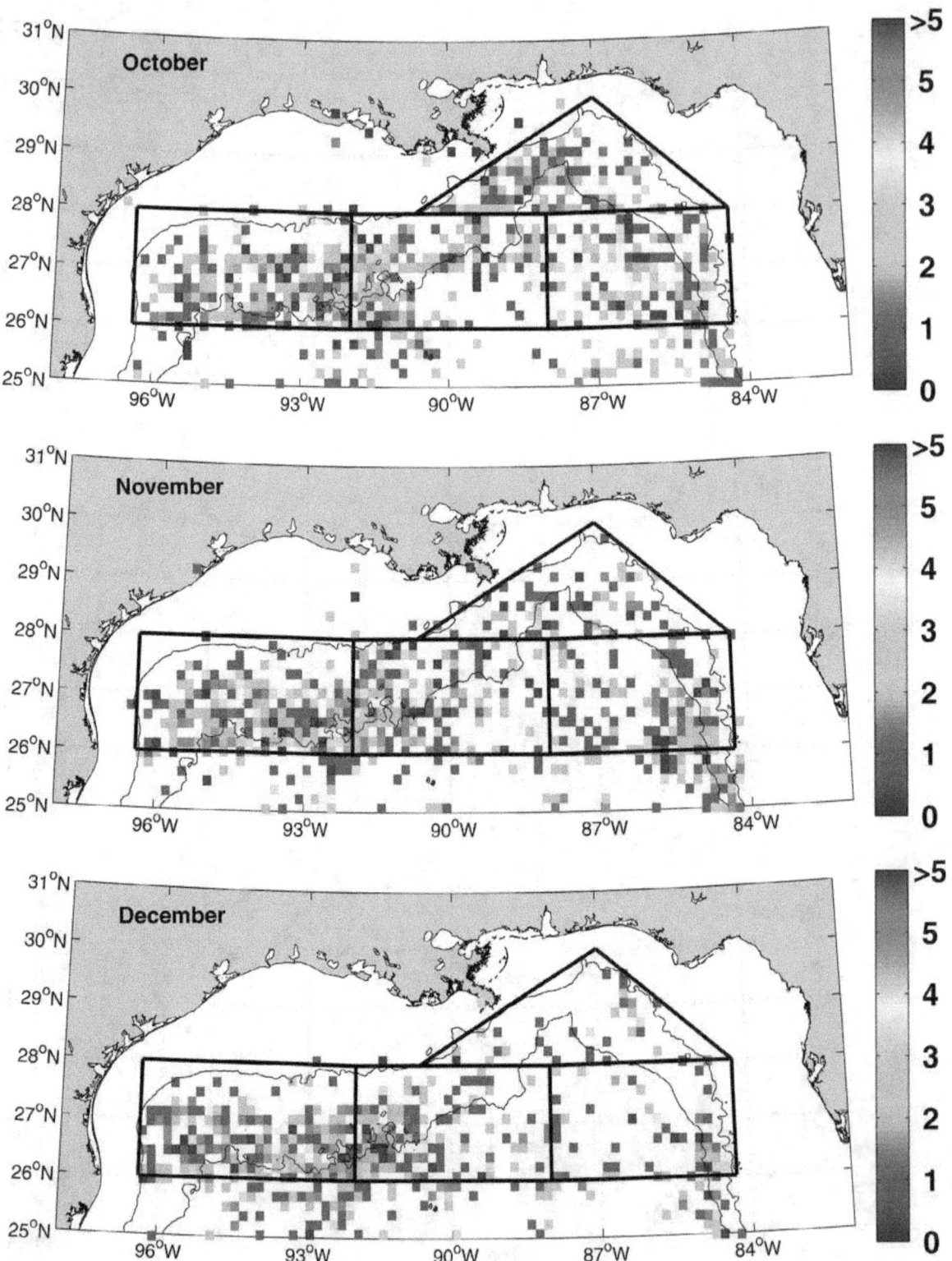

Figure 36. Catch per unit effort of adult dolphinfish from the commercial long-line fishery. Each square represents the mean catch-per-unit-effort (dolphinfish per set) taken within a 10' x10' region from January through December over the period 1986-1999. Note that the colorbar was arbitrarily limited to a maximum CPUE of 5. The maximum CPUEs were: January =27; February =27; March=109; April =70; May =126; June=168.5; July =132.3; August=100; September=132.5; October = 38; November = 67; December=22. (continued)

4.6.2 Reproduction

Both sexes of dolphinfish can reach sexual maturity in their first year of life. Female dolphin can reach sexual maturity at 350 mm FL and 100% of females at 550 mm FL were mature, while male dolphin began to reach sexual maturity at 427 mm FL (Beardsley, 1967).

Examination of gonadal development in female dolphinfish in the Florida Current suggested an extended spawning period from January to October with peak spawning from January through March (Beardsley, 1967). The year-round presence of juvenile dolphinfish in the Florida Current suggests that spawning may continue throughout the year (Beardsley, 1967) although a summer peak in juvenile abundance supports the hypothesis of intensive spawning during the first quarter of the year. Gibbs and Collette (1959) examined size distributions of *C. hippurus* from the Gulf of Mexico and these data suggested an extended spawning period from April through July or August. Ditty et al. (1994) found *C. hippurus* larvae present throughout the year in the northern Gulf of Mexico with small larvae more abundant during the summer months. Ditty et al. (1994) believed that spawning occurs in oceanic waters because most preflexion larvae were collected at stations beyond the continental shelf.

Eggs are buoyant and hatch within 60 h of fertilization at 24-25 °C (Palko et al, 1982). Newly hatched larvae are approximately 3 mm and growth is rapid with larvae reaching 15 mm at 15 days of age (Hassler and Rainville, 1975). Larvae become juveniles at 10-20 mm SL (Palko et al., 1982).

4.6.3 Larval/Juvenile Distributions

Preflexion larvae occurred from April through November and over 90% of larvae occurred in samples collected over the outer continental shelf and in oceanic waters (Ditty et al. 1994). Ditty et al. (1988) found larval *C. hippurus* present in the northern Gulf of Mexico from February through November. In a later study using the SEAMAP dataset, they found a relationship between larval presence and temperature and salinity. Over 90% of larvae were collected when water temperatures were 24 °C or greater and few larvae were collected below 25 psu and most occurred above 32 psu (Ditty et al., 1994).

Larval *C. hippurus* were present during all months of the year in the northern Gulf of Mexico (Fig. 37). Relatively few stations yielded larvae from January through March and larvae were primarily found well offshore. From April through June, larvae were more frequently detected within survey areas and were found further to the north and over the shelf waters. It was difficult to assess their distributions in the study area during summer, however, larvae were common along the northern edge of the study zones and were found over the slope water during August (Fig. 37) suggesting that they are also likely present there during July and September. The distributions of larvae in the study area during October and November could not be assessed due to limited sampling coverage. In December, larvae were present in the southern part of the study area in oceanic and mid-depth slope waters of the eastern, central, southern portion of the northern zone and on the eastern edge of the western zone (Fig. 37).

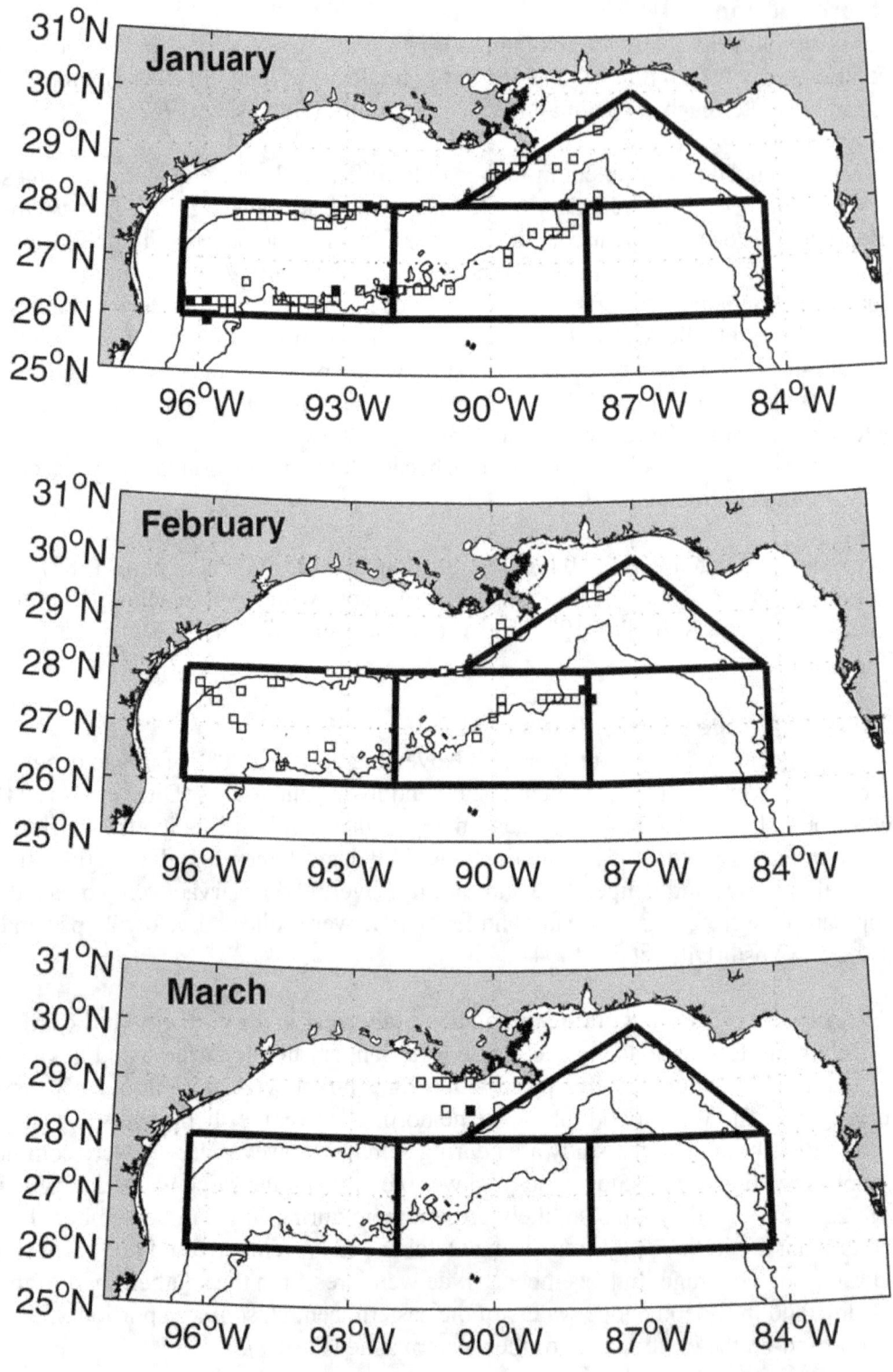

Figure 37. Presence (■) and absence (□) of dolphinfish larvae in the study area from January through December estimated from SEAMAP ichthyoplankton data.

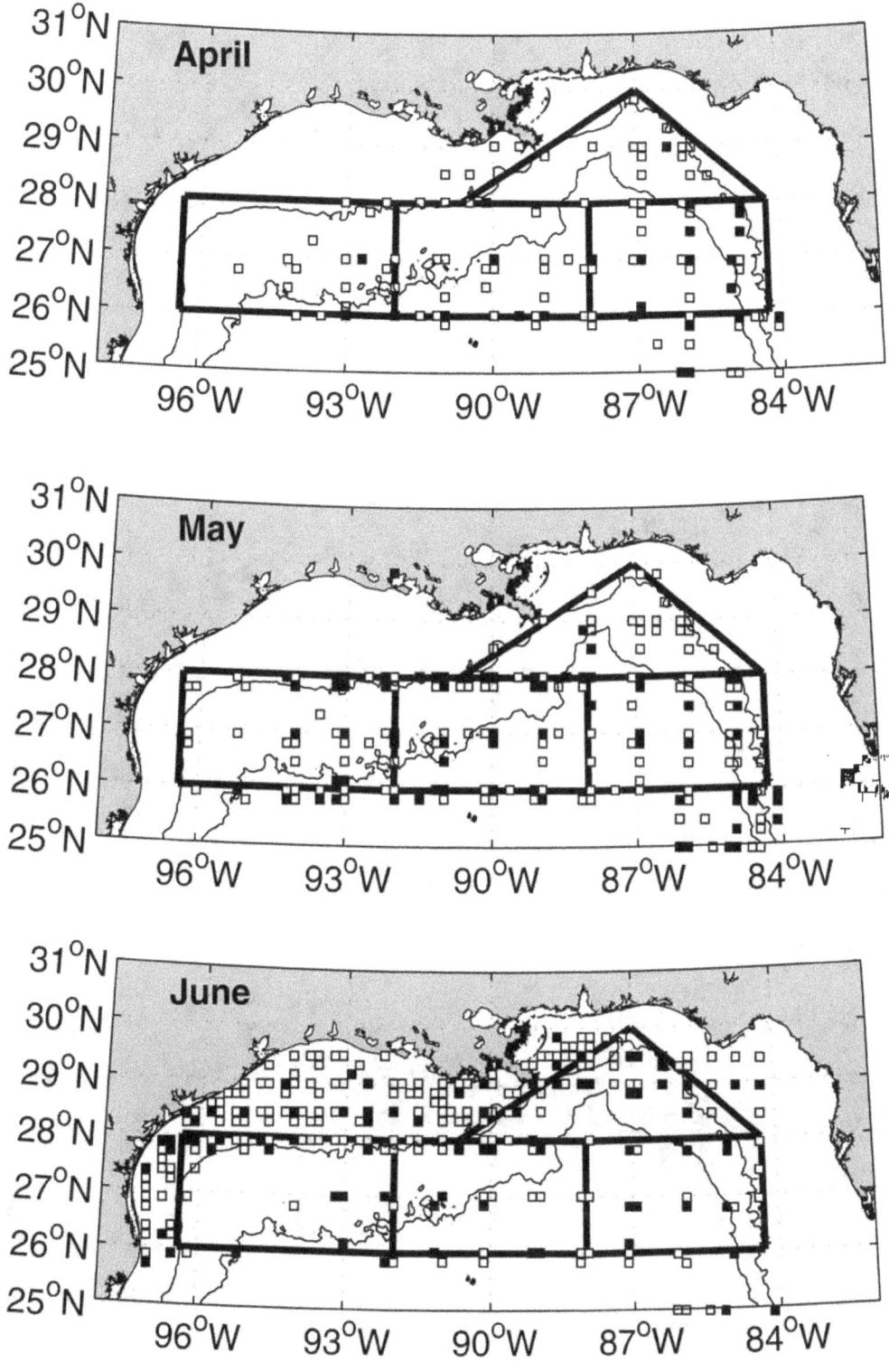

Figure 37. Presence (■) and absence (□) of dolphinfish larvae in the study area from January through December estimated from SEAMAP ichthyoplankton data. (continued)

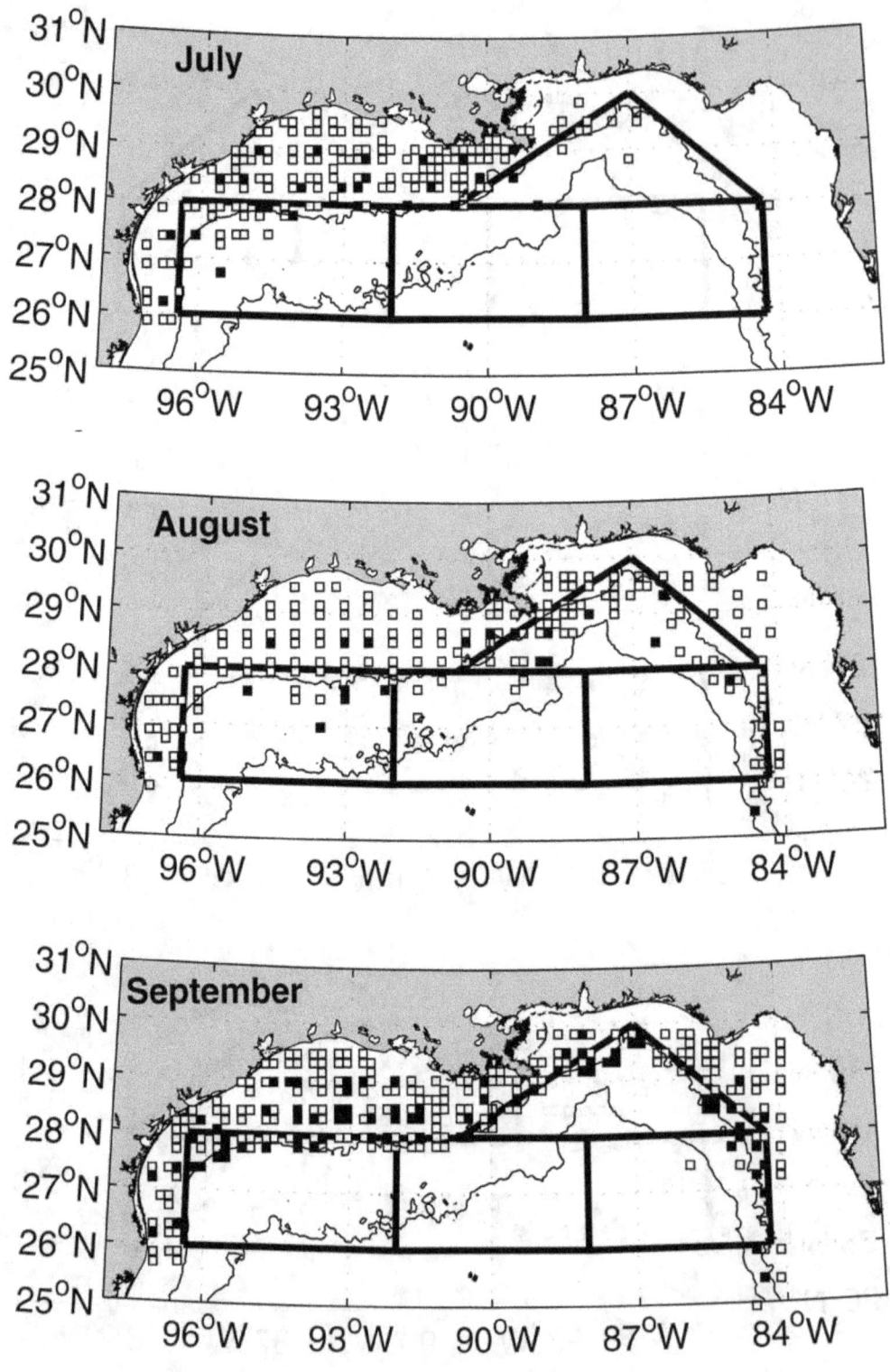

Figure 37. Presence (■) and absence (□) of dolphinfish larvae in the study area from January through December estimated from SEAMAP ichthyoplankton data. (continued)

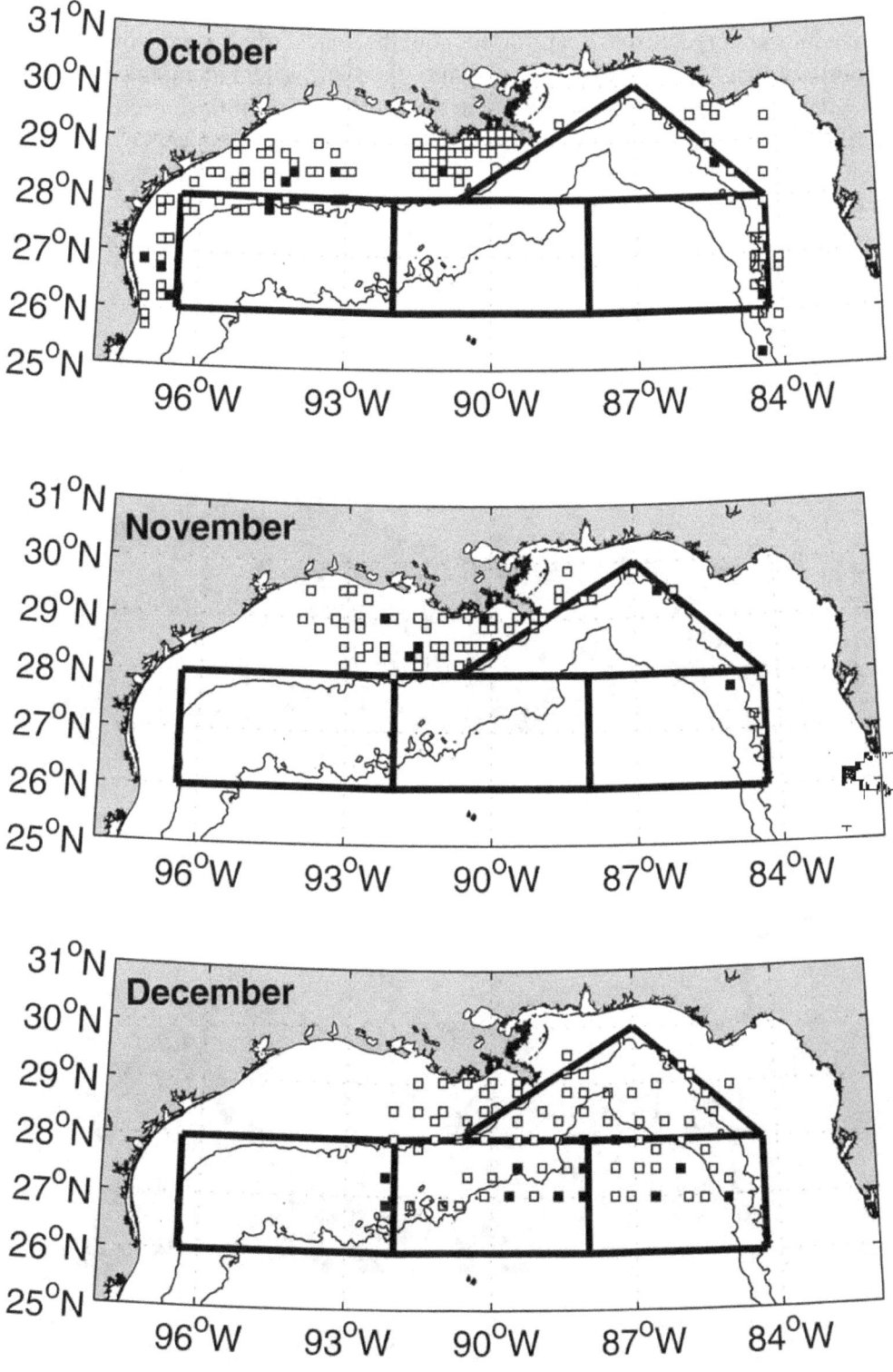

Figure 37. Presence (■) and absence (□) of dolphinfish larvae in the study area from January through December estimated from SEAMAP ichthyoplankton data. (continued)

4.6.4 Predicted Adult Distributions

Dolphinfish adults are predicted to occur throughout the waters of all study zones during January, February, and March with the exception of the shelf waters along the eastern edge of the eastern zone during January (Fig. 38). A void in the southern part of the eastern zone expands during March to cover a large section of the oceanic waters. This disappears by April. From May through November, dolphinfish are present throughout the study area with the intermittent exception of the northwestern corner of the northern zone (Fig. 38). During December they are present throughout the study area except for the shelf waters south of Mississippi and the extreme northwest corner of the western zone (Fig. 38).

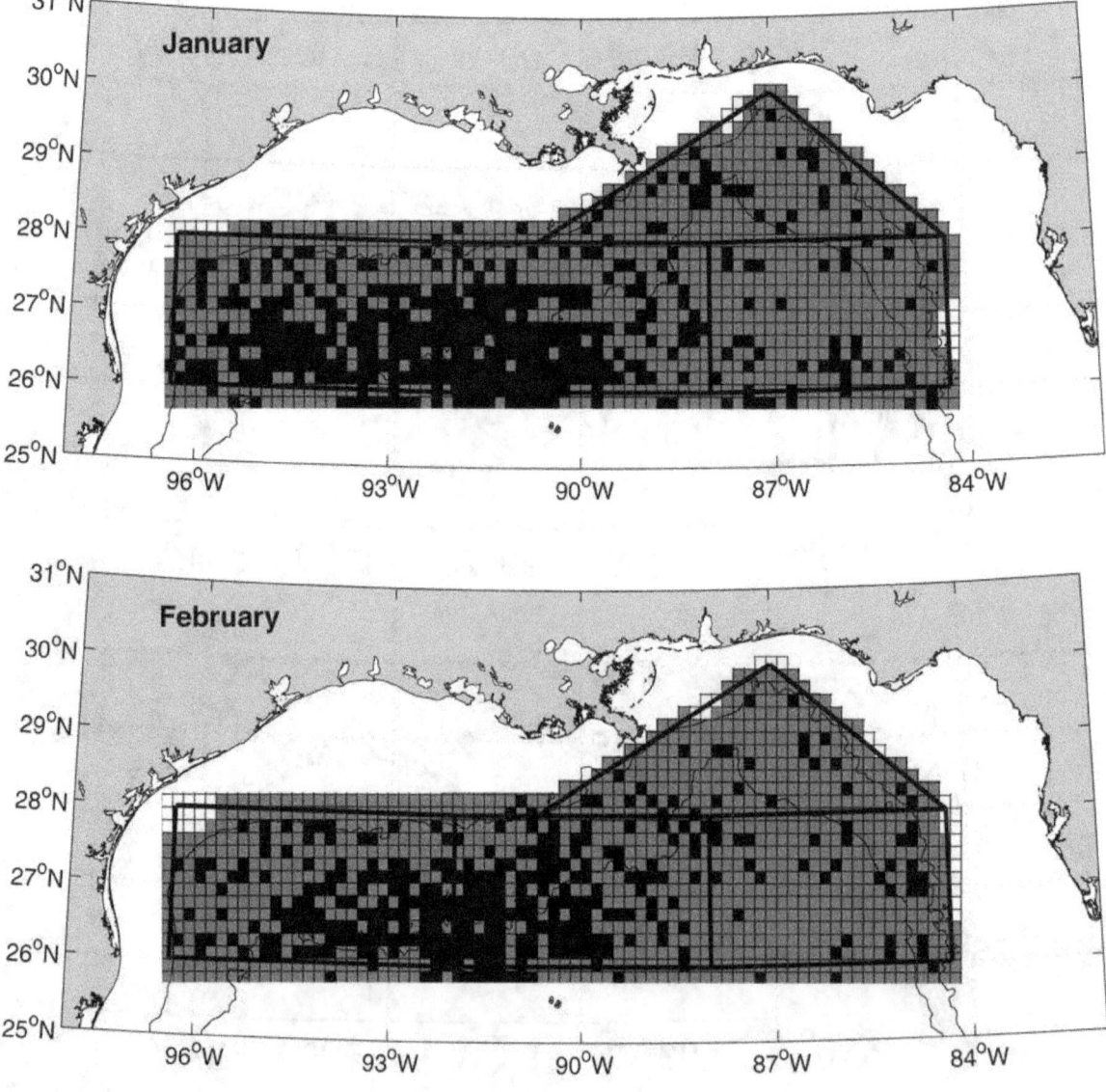

Figure 38. Predicted distributions of adult dolphinfish in the study area from January through December. The presence of individuals in each grid cell is coded as: confirmed (■), reasonable inference (▨), or unreported (□).

110

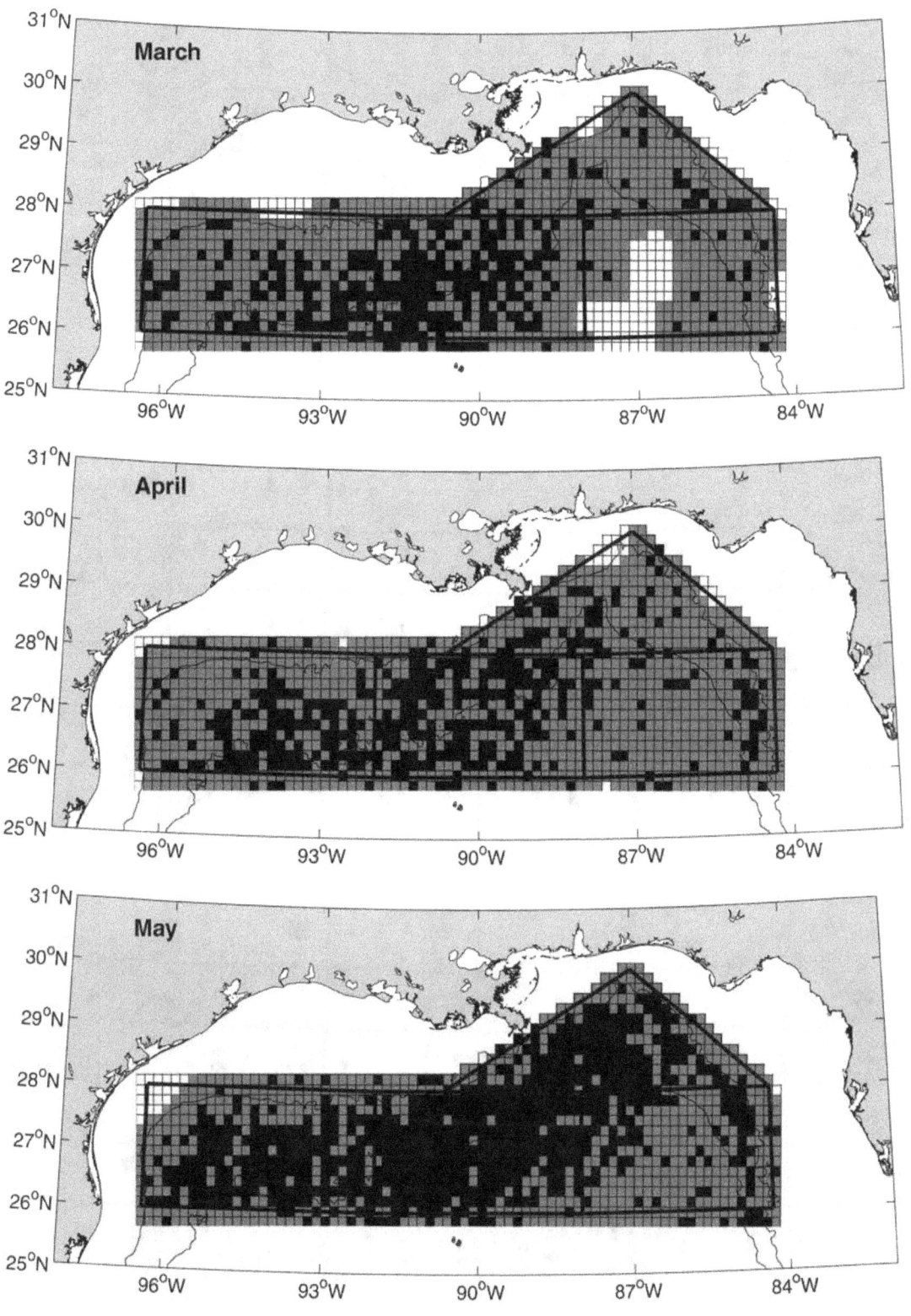

Figure 38. Predicted distributions of adult dolphinfish in the study area from January through December. The presence of individuals in each grid cell is coded as: confirmed (■), reasonable inference (▧), or unreported (□). (continued)

111

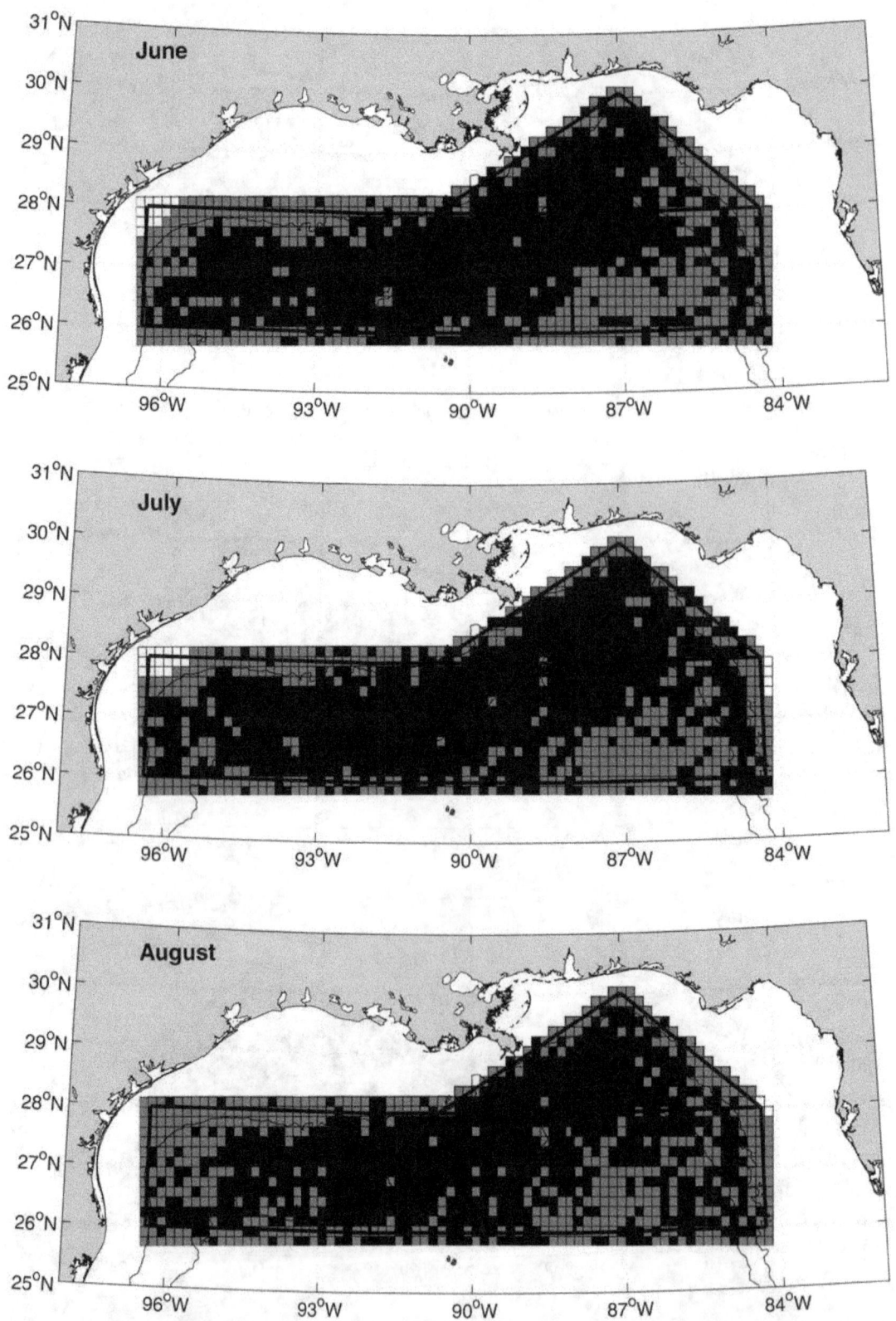

Figure 38. Predicted distributions of adult dolphinfish in the study area from January through December. The presence of individuals in each grid cell is coded as: confirmed (■), reasonable inference (▨), or unreported (□). (continued)

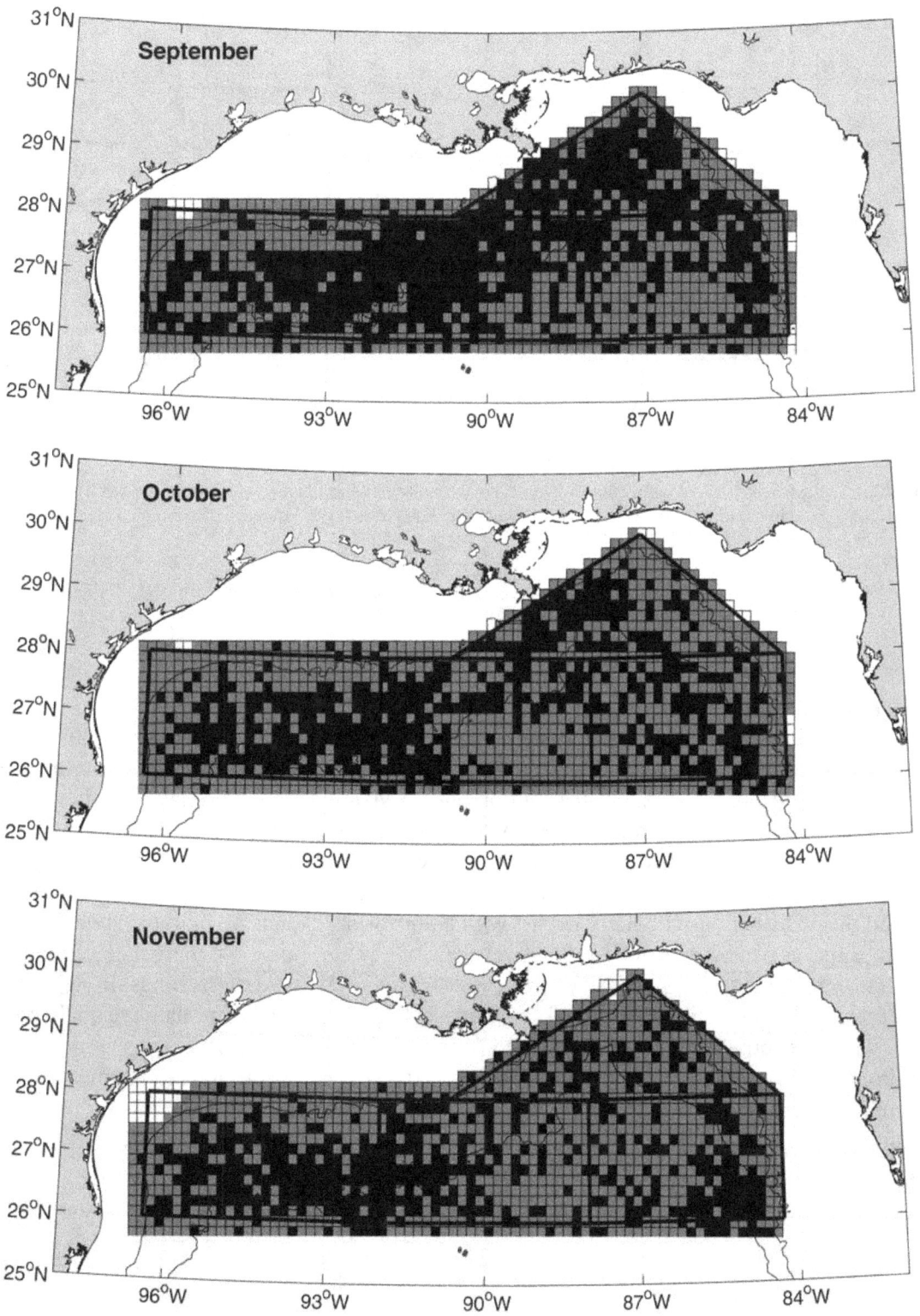

Figure 38. Predicted distributions of adult dolphinfish in the study area from January through December. The presence of individuals in each grid cell is coded as: confirmed (■), reasonable inference (▨), or unreported (□). (continued)

113

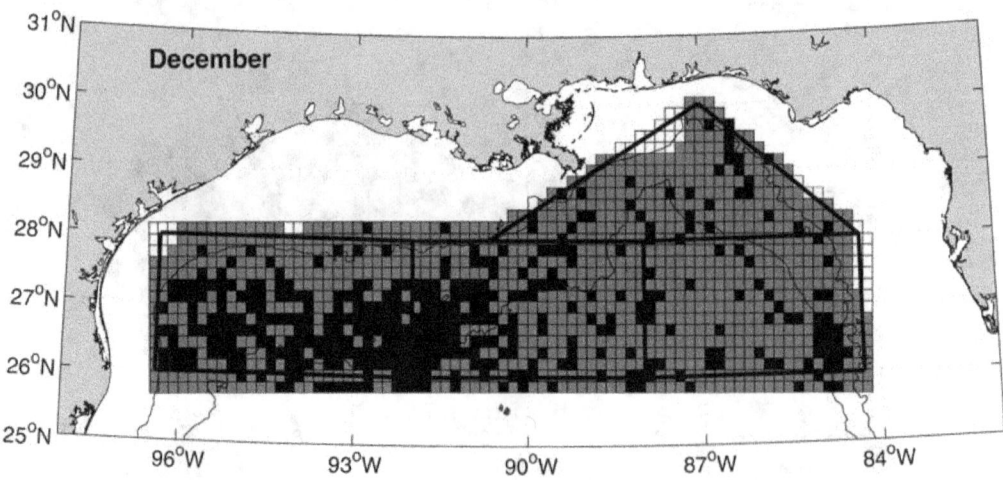

Figure 38. Predicted distributions of adult dolphinfish in the study area from January through December. The presence of individuals in each grid cell is coded as: confirmed (■), reasonable inference (▨), or unreported (□). (continued)

4.6.5 Predicted Larval/Juvenile Distributions

Larvae are predicted to be present in the study zone throughout the year. During January and February they will be present along the southern edge of the slope water and likely in the oceanic waters to the south (Fig. 39). Their distribution during March cannot be reliably predicted in specific cells, but it is likely similar to that of April when larvae are present in the oceanic and slope water regions of the central, eastern, and the southerly part of the northern zone (Fig. 39). During May they will be present over the shelf edge, slope water and oceanic waters of the study zone and this distribution will likely persist with a northward expansion onto the shelf during June and July. In August they are present over the slope water, however, there are no sample records to confirm their likely presence in the western, central and eastern zones. In September larvae are very common along the shelf-slope break (Fig. 39). Insufficient data are available to predict their distributions during October and November; however, cooling water in the northern Gulf probably stimulates southerly migration and shifts spawning activities to warmer water in the southern parts of the study area. In December, larvae are present over the mid-slope and oceanic waters of the central and eastern zones and are likely in comparable waters within the western zone as well. The distribution of juveniles likely overlaps that of the larvae.

114

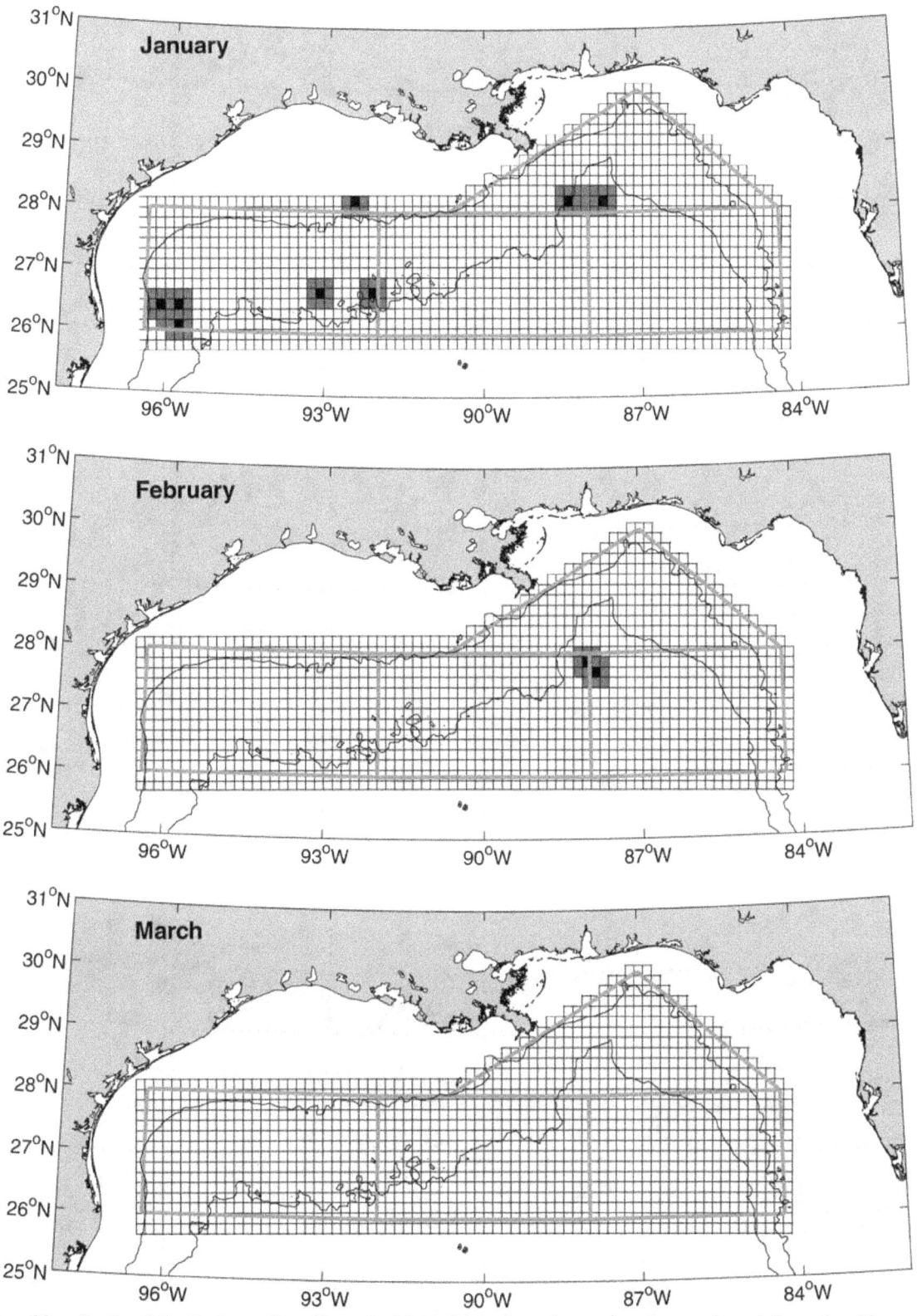

Figure 39. Predicted distributions of larval/juvenile dolphinfish in the study area from January through December. The presence of individuals in each grid cell is indicated as confirmed (■), reasonable inference (■) or unreported (□).

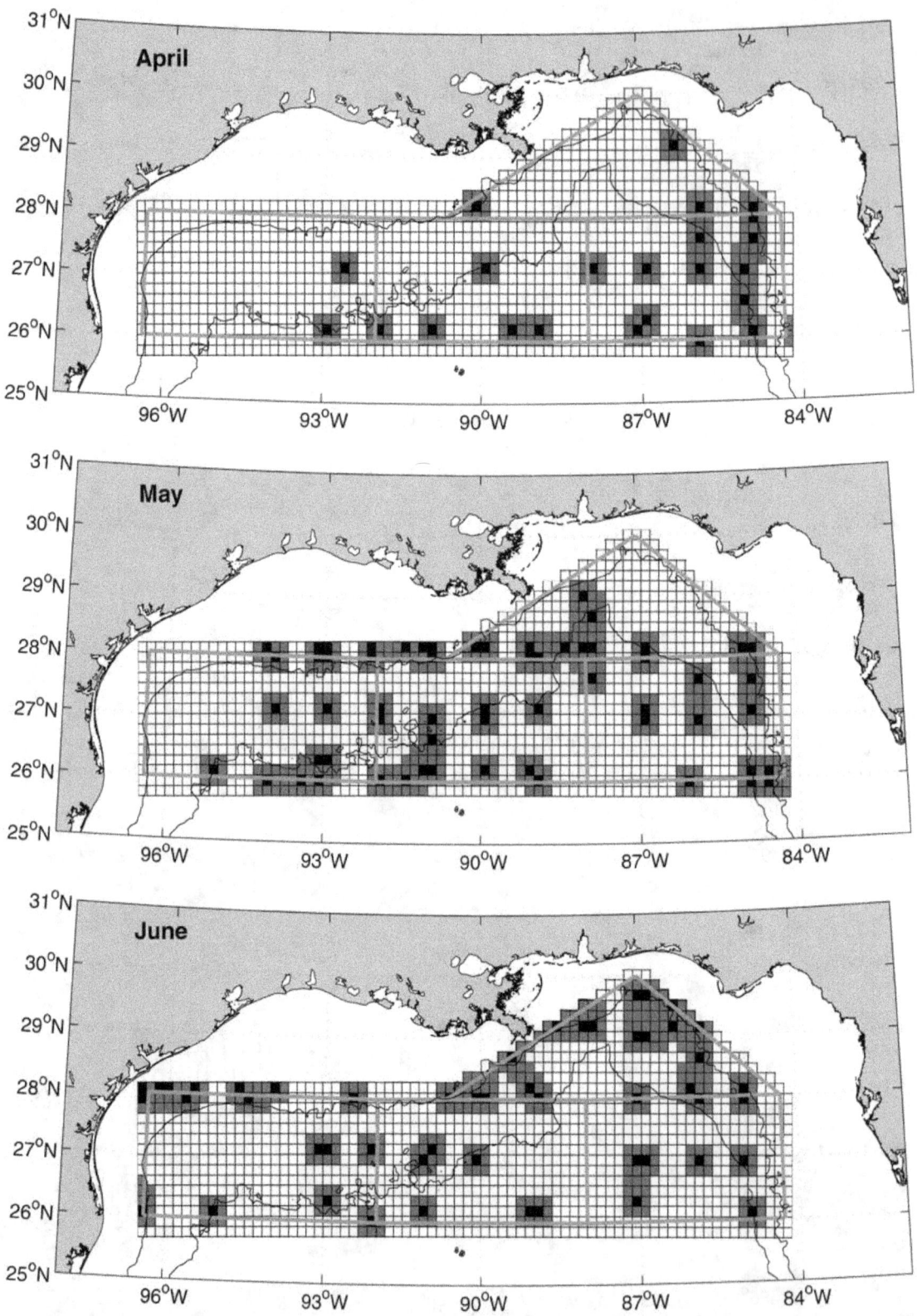

Figure 39. Predicted distributions of larval/juvenile dolphinfish in the study area from January through December. The presence of individuals in each grid cell is indicated as confirmed (■), reasonable inference (■) or unreported (□). (continued)

116

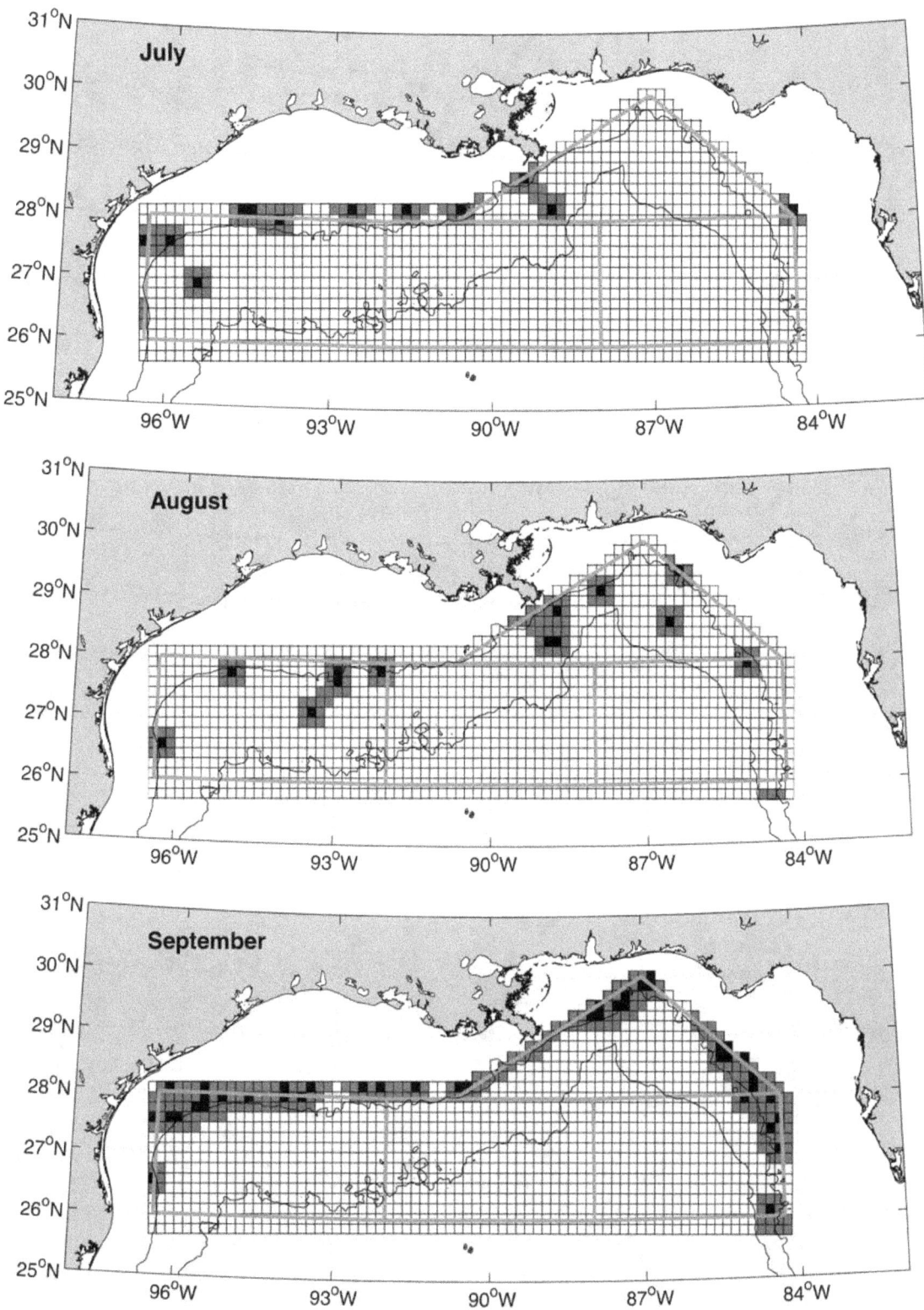

Figure 39. Predicted distributions of larval/juvenile dolphinfish in the study area from January through December. The presence of individuals in each grid cell is indicated as confirmed (■), reasonable inference (▨) or unreported (□). (continued)

117

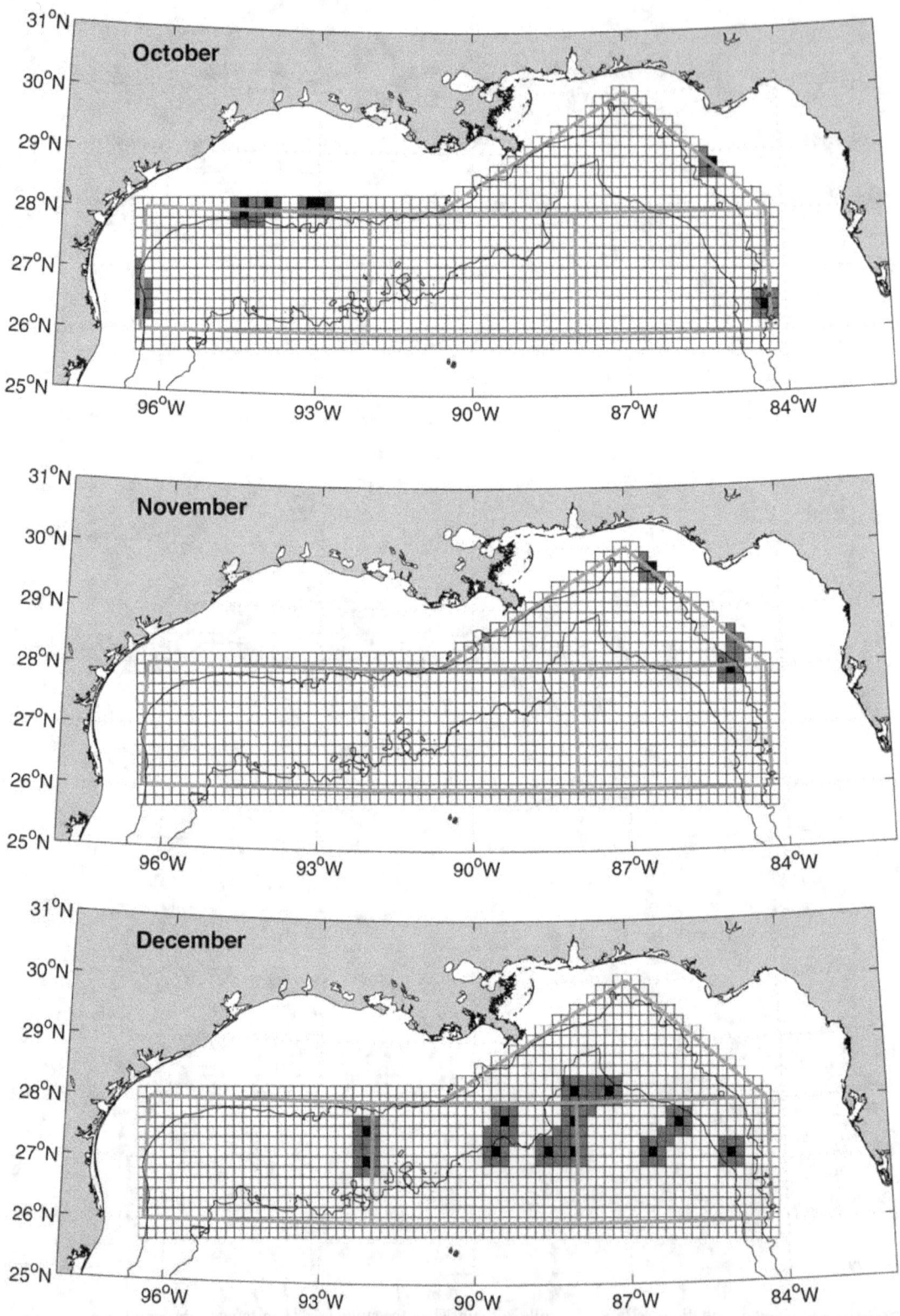

Figure 39. Predicted distributions of larval/juvenile dolphinfish in the study area from January through December. The presence of individuals in each grid cell is indicated as confirmed (■), reasonable inference (■) or unreported (□). (continued)

4.7   Blue Runner (*Caranx crysos*)

Blue runner are important forage fish for larger pelagic predators. Dense schools of blue runner are frequently associated with petroleum platforms and this species is a component of recreational trolling landings in the Gulf of Mexico. A survey of landings reported by U.S. charterboat captains in the southeastern U.S. and Caribbean including the Gulf of Mexico indicated that approximately 2,735 blue runner were collected by trolling and this number was 5.1% of the total catch (Brusher and Palko, 1987). In that same survey, blue runner ranked among the most abundant species collected by trolling off: northwest Florida (1984: 638 fish, 28.3%; 1985: 443 fish, 26.9%); Alabama (1984: no data; 1985: 61 fish; 4.4%); Mississippi (1984: no data; 1985: 210 fish; 3.6%); Louisiana (1984: 79 fish, 3.5%; 1985: 183 fish, 3.0%); north Texas (1984: no data; 1985: 84 fish; 1.6%); and south Texas (1984: 68 fish, 4.1%; 1985: 173 fish, 4.7%). A limited commercial fishery exists in inshore waters outside of the boundaries of the present study area during summer (NOS, 1985).

4.7.1 Adult Distributions

Blue runner are commonly associated with offshore petroleum platforms (Keenan et al. 2003). Stanley and Wilson (1997) recorded their presence year round at an inner shelf platform (WC352, 28.9892°, -93.5058°) with a peak in abundance during summer and lowest densities during winter. NOS (1985) provides an annual distribution map that suggests that their range encompasses all of the present study area. Patillo et al. (1997) indicate that blue runner occur seasonally throughout the Gulf of Mexico with high densities off the west coast of Florida and south of Mississippi. NOS (1985) reports that they are distributed throughout the study area year-round (Fig. 40).

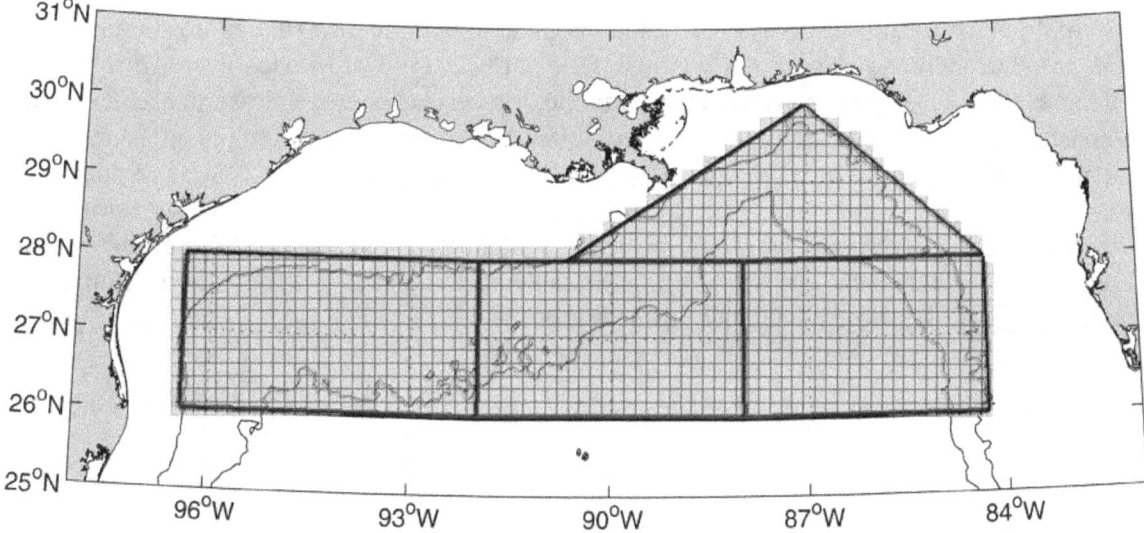

Figure 40.   Year round distribution of blue runner (■) in the study area based on their distribution in NOS (1985).

4.7.2 Reproduction

Spawning occurs during July and August (Christmas et al., 1974). Ditty et al. (1988) reported the presence of larvae from March through November with peak abundances during June, July and August. This suggests some spawning from spring through fall with a summer peak. Dooley (1972) felt that based on cohort analysis, spawning occurred primarily during early September in the Florida Current. Goodwin and Finucane (1985) examined gonads from blue runner collected

119

off south Florida, northwest Florida and off the Mississippi Delta by commercial fisheries. Their results indicated that maximal spawning occurred from June through August for fishes in the northern Gulf of Mexico with a secondary peak in spawning during October off northwest Florida.

4.7.3 Larval/Juvenile Distributions

Larvae of *C. crysos* were present in the northern Gulf of Mexico from March through November with peak abundances during June, July and August (Ditty et al., 1988). Larvae were reported to be common in the upper water column (<40 m) during summer with records to depths as great as 182 m (LGL and SAIC, 1993). Most larvae reported in this study were collected in waters where the maximum depth ranged from 50-200 m and Patillo et al. (1997) indicate that larvae are most abundant over the continental shelf. A review of the distributions of carangid larvae in the eastern Gulf using bongo nets (Leak, 1981) indicated that *C. crysos* larvae were present in waters from 46-1646m depth at salinities from 31-36 psu and temperatures from 29-30 °C. During that study blue runner were only sampled during August, which accounts for the rather warm and limited temperature range.

Blue runner become juveniles at approximately 12 mm SL (McKenny et al. 1958) and transition to sexually mature adults somewhere between 225 mm SL (males) and 247 mm SL (females) (Patillo et al. 1997). Collections from neuston nets in the eastern Gulf detected juvenile *C. crysos* during May through July over depths of 70-225 m (Leak, 1981). Juveniles have been reported across the continental shelf waters of the northern Gulf of Mexico west of the Mississippi Delta (Shaw and Drullinger, 1990). Young (juvenile) blue runner are associated with *Sargassum* (Dooley, 1972; Christmas et al. 1974) and other flotsam (Patillo et al. 1997).

SEAMAP ichthyoplankton surveys reveal that larval and juvenile blue runner are abundant throughout the shelf and slope water regions of the northern Gulf of Mexico from April-September (Fig. 41). Sampling coverage within the deepwater zones was less intense than within the coastal waters, however, the samples suggest that blue runner larvae and juveniles are uncommon in the western area, and present in the central, northeastern and eastern regions during April, May and June (Fig. 41). The abundance of larval and juvenile blue runner in the northeastern area appears to increase from July to September, however, sparse sampling within the western, central and eastern areas during the same period makes it difficult to determine whether they are also more abundant in these regions as well.

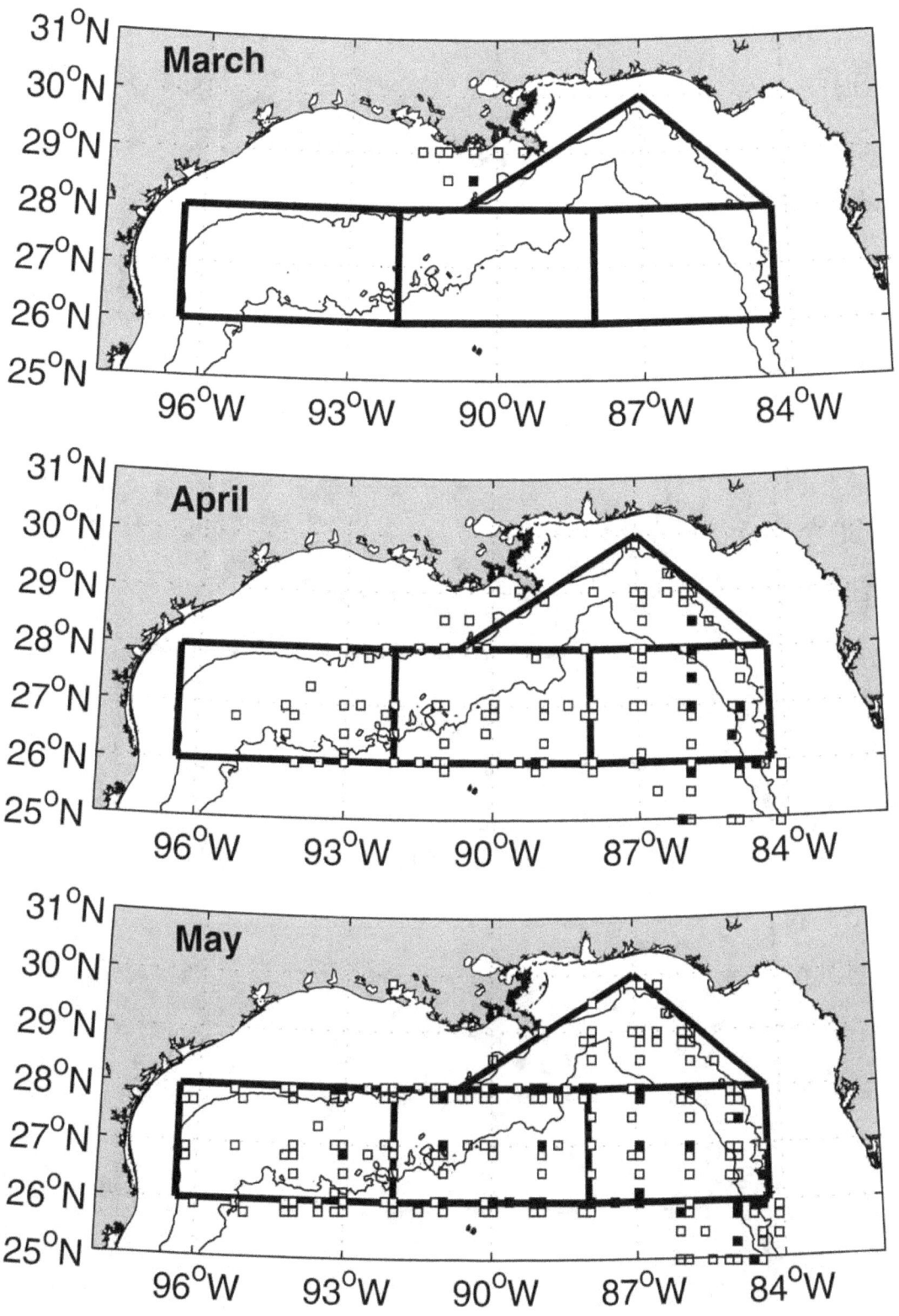

Figure 41. Presence (■) and absence (□) of blue runner larvae in the study area from March through December estimated from SEAMAP ichthyoplankton and other sampling data.

121

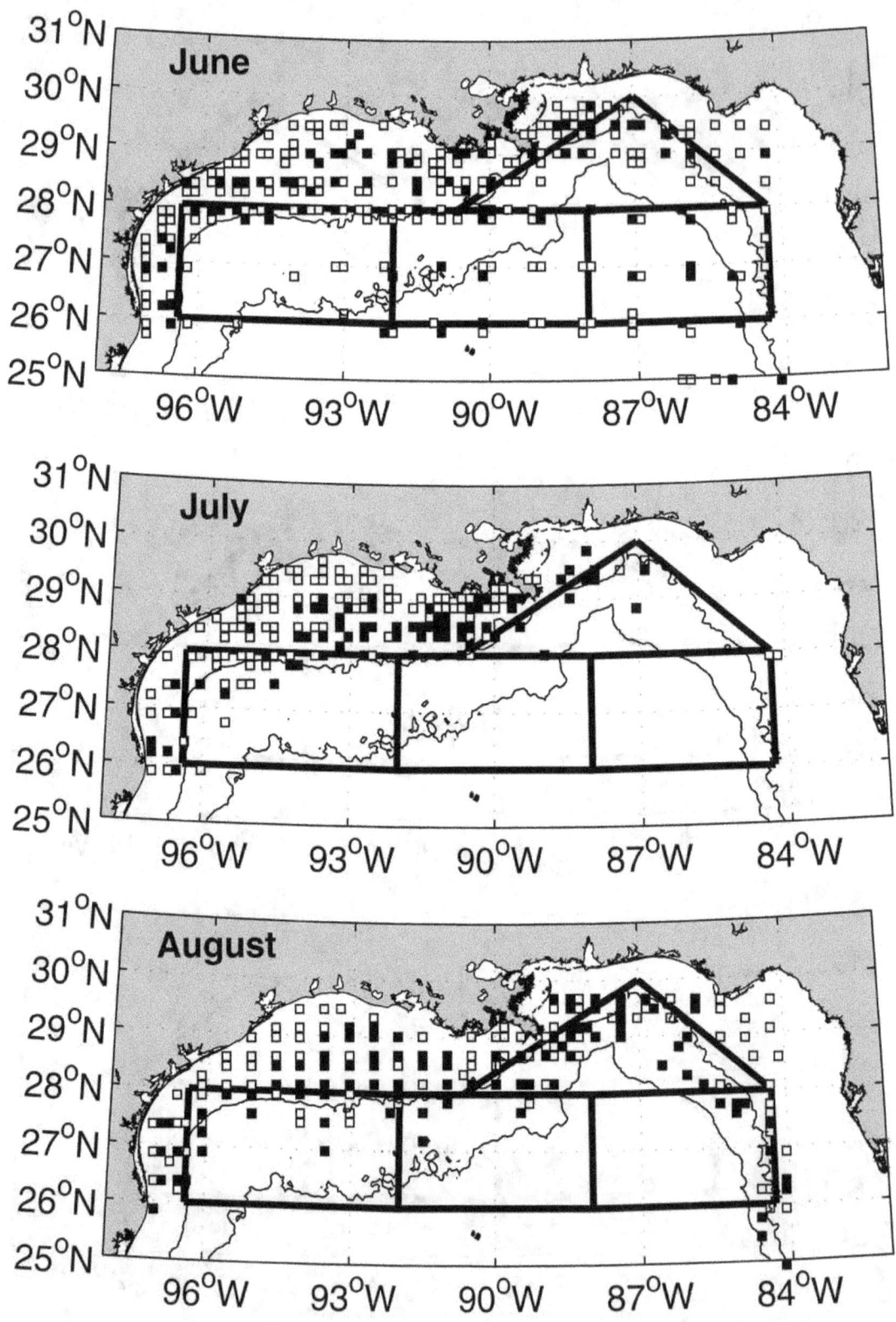

Figure 41. Presence (■) and absence (□) of blue runner larvae in the study area from March through December estimated from SEAMAP ichthyoplankton and other sampling data. (continued)

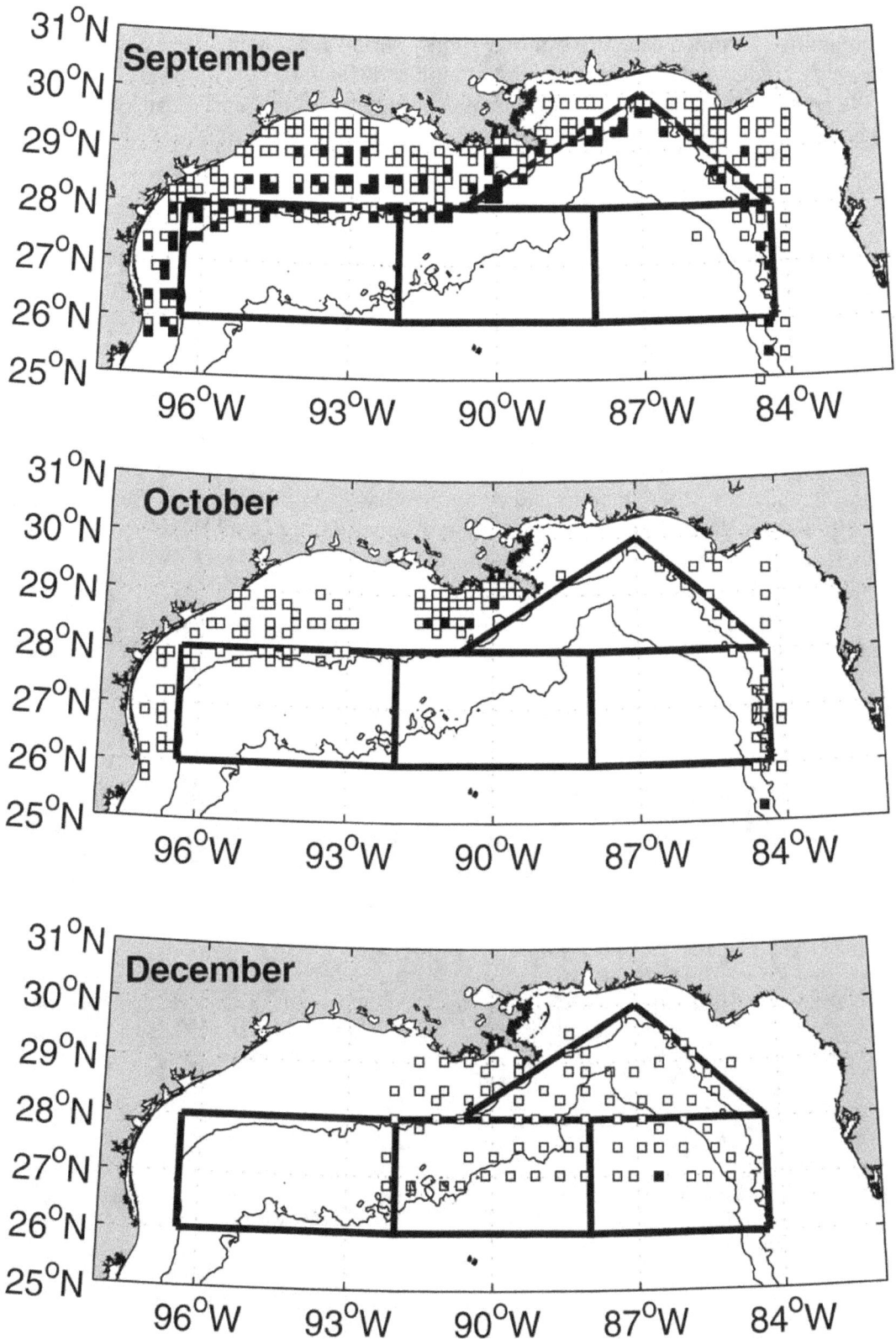

Figure 41. Presence (■) and absence (□) of blue runner larvae in the study area from March through December estimated from SEAMAP ichthyoplankton and other sampling data. (continued)

4.7.4 Predicted Adult Distributions

Data containing confirmed detections of blue runner during each month were not available, and thus, monthly predictions of their probable spatial locations in the study area are not possible. Given the reported broad distribution of blue runner in the northern Gulf of Mexico by NOS (1985) and others, it is likely that blue runner are present throughout the study area year round.

4.7.5 Predicted Larval/Juvenile Distributions

Larval blue runner will be present in the study area from April through December. During April they will be distributed throughout the eastern halves of the eastern and northern zone and scattered in the central zone (Fig. 42). Though not detected in the western zone by SEAMAP, it is probable that they are present in that area as well. By May, larvae will be found throughout the eastern and central zones and along the southern edge of the northern zone. Larvae are likely present throughout the western zone as well. In June, larvae will be common along the shelf-slope break in the northern, central and western zones and scattered over deeper waters throughout the study area (Fig. 42). This pattern will continue through July, August, and September. Though detected outside of the study area during October and November, there were no records inside during these months and consequently no spatial prediction. Scattered larvae are likely present in the western, central and eastern zones during October, November and December. Juveniles will co-occur with larvae.

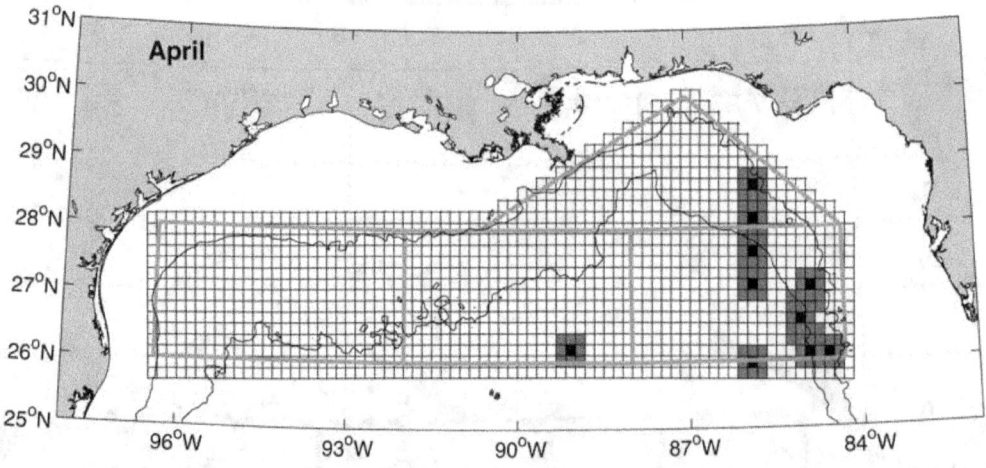

Figure 42. Predicted distributions of larval/juvenile blue runner in the study area from April through September and December. The presence of individuals in each grid cell is indicated as confirmed (■), reasonable inference (▨) or unreported (□).

124

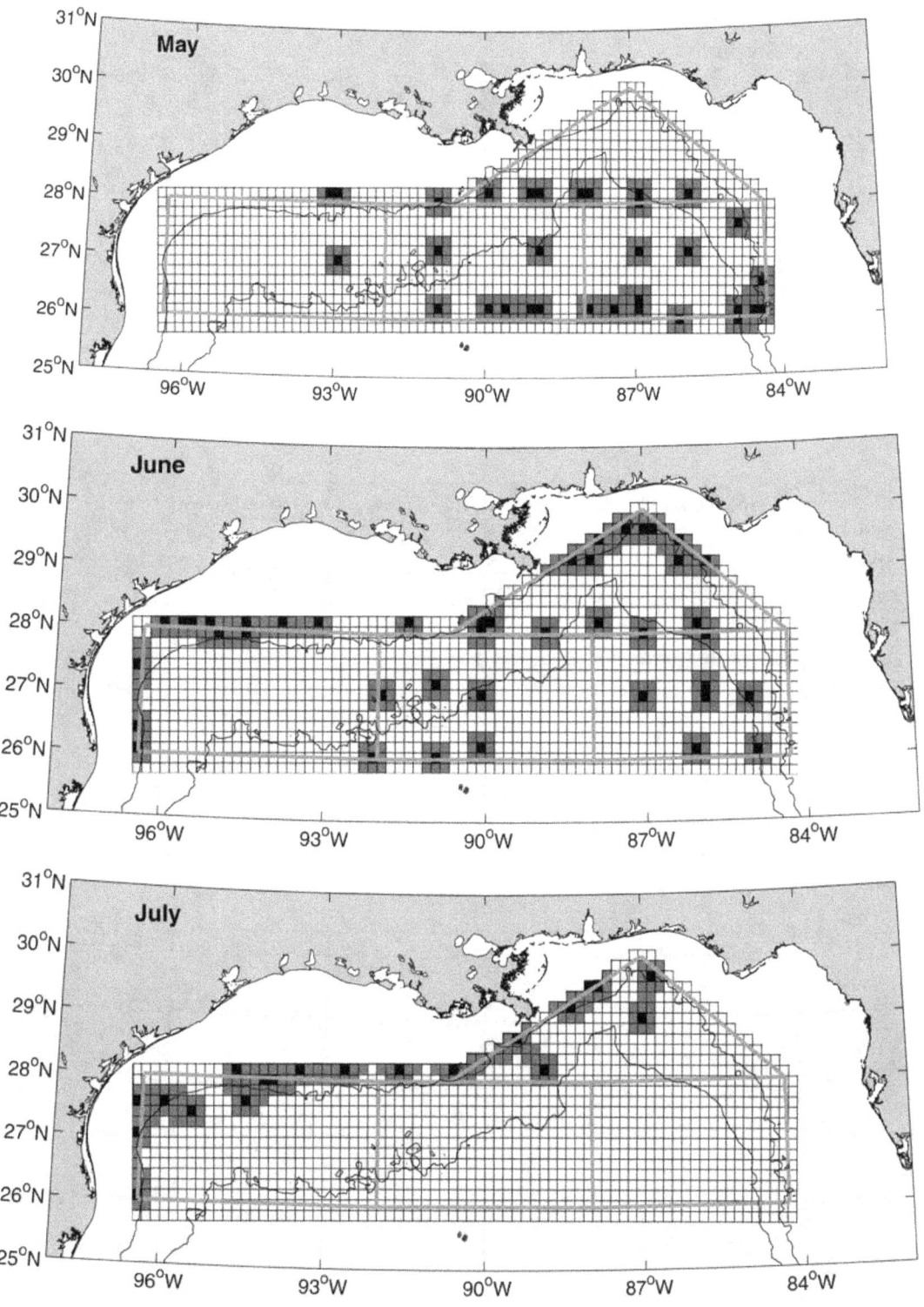

Figure 42. Predicted distributions of larval/juvenile blue runner in the study area from April through September and December. The presence of individuals in each grid cell is indicated as confirmed (■), reasonable inference (▦) or unreported (□). (continued)

125

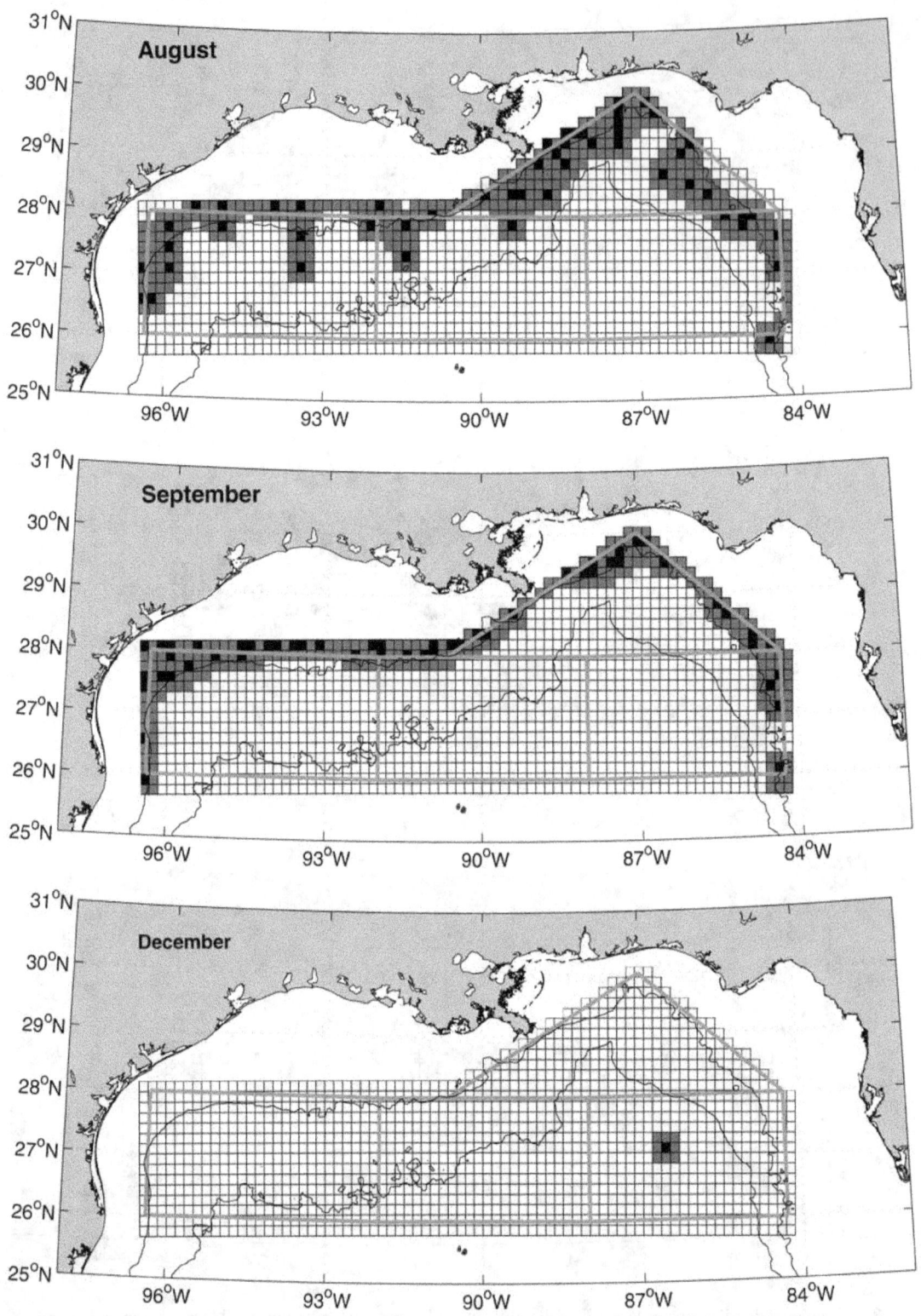

Figure 42.   Predicted distributions of larval/juvenile blue runner in the study area from April through September and December. The presence of individuals in each grid cell is indicated as confirmed (■), reasonable inference (▦) or unreported (□). (continued)

126

4.8   Flyingfishes

Flyingfishes are frequently associated with floating *Sargassum* (South Atlantic Fishery Management Council, 1998; Table 4). The eggs, larvae, juveniles and adults of two common Gulf of Mexico species: the spotfin flyingfish *Cypselurus furcatus* and the Atlantic flyingfish *C. melanurus* (previously *C. heterurus*) have been found in association with floating *Sargassum* in South Atlantic waters.

Adult *C. melanurus* are found in tropical and temperate waters and are common in the Gulf of Mexico (Staiger 1965). This species is generally confined to coastal waters within 240 km of land and juveniles are frequently found very close to shore (Staiger 1965). Reproductive seasonality in the Gulf of Mexico is unknown, however, the species spawns from June to July in Atlantic waters off Morocco and from May to August in the Mediterranean.

Distributional data on flyingfishes were provided by Dr. Robert L. Pitman (NMFS Southwest Fisheries Science Center, La Jolla, California). Those data consisted of *C. furcatus* and *C. melanurus* collected at various locations (Fig. 43) in the Gulf of Mexico during a series of research cruises during April through June between 1990 and 1996. The database also contains a few additional observations from other months that were not part of a systematic survey. Specimens were collected via dipnet and other methods. This sampling approach appears to collect juveniles of approximately 20 mm fork length (FL) through small adults of up to 270 mm FL were also collected. Coverage within the study area during April was fair in the central, eastern and northern zones (Fig. 43). During May there was good coverage throughout all four zones, and during June there was sparse coverage in each zone (Fig. 43).

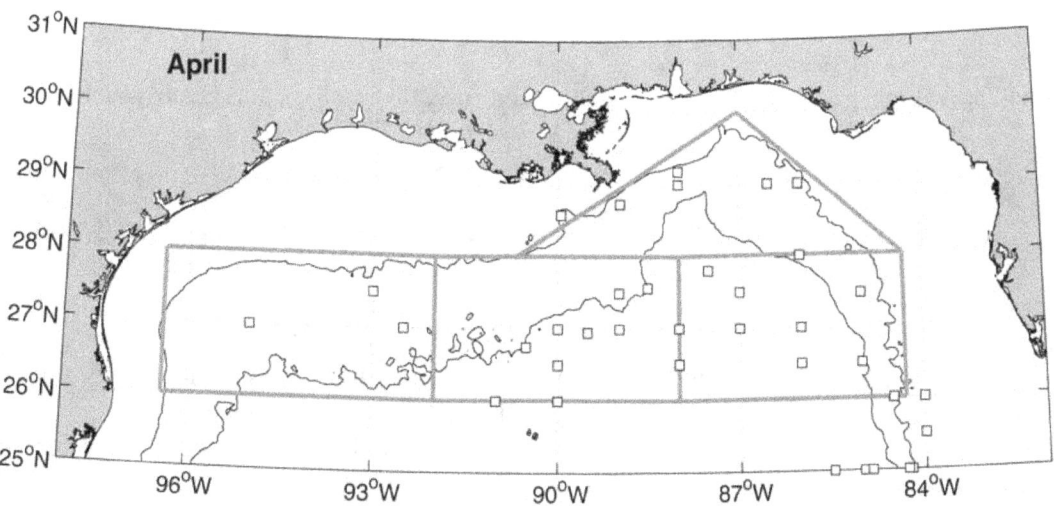

Figure 43.   Locations (□) where sampling for flyingfishes was conducted by NOAA (Dr. R.L. Pitman, pers comm.) in the Gulf of Mexico from April through June over the period 1990-1996.

127

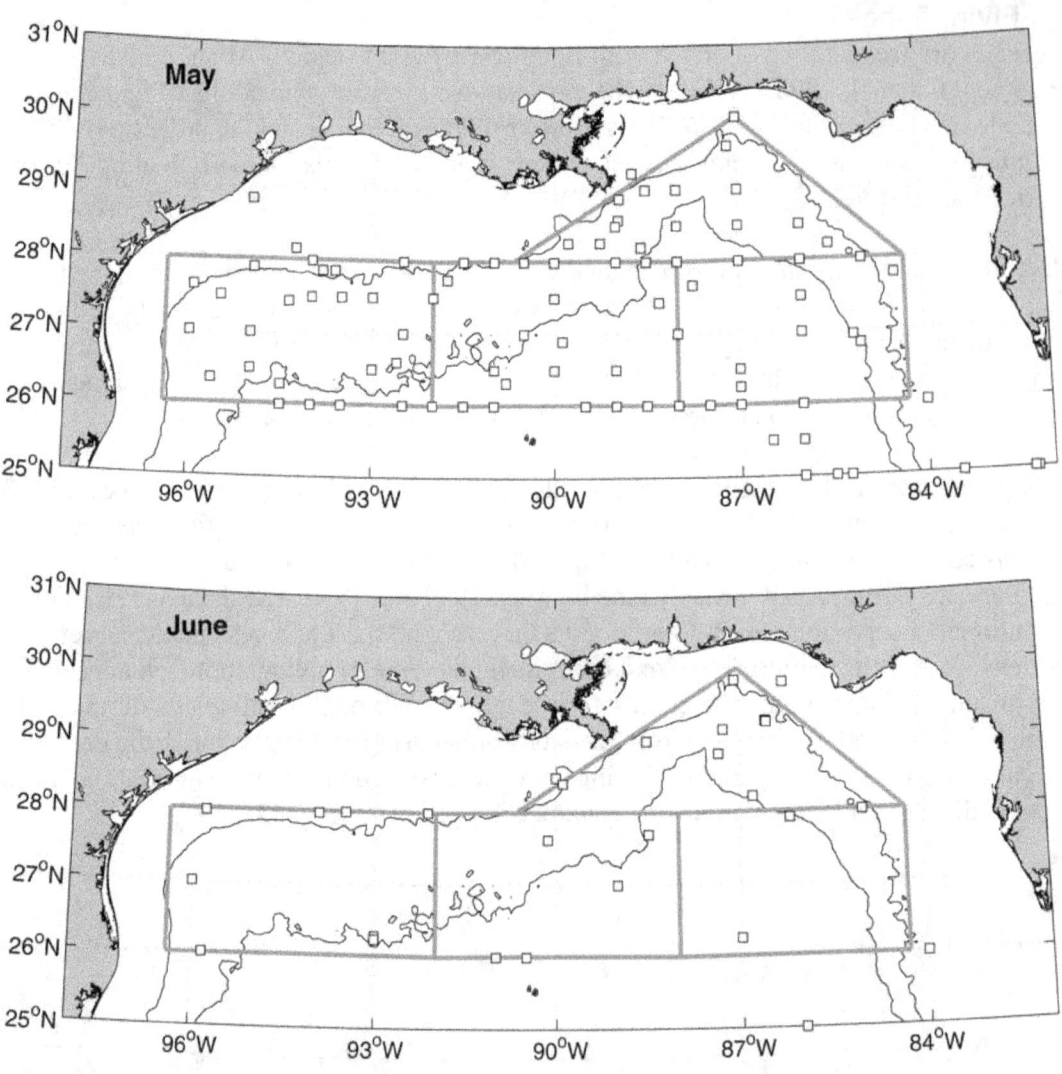

Figure 43. Locations (□) where sampling for flyingfishes was conducted by NOAA (Dr. R.L. Pitman, pers comm.) in the Gulf of Mexico from April through June over the period 1990-1996. (continued)

4.8.1 Reproduction

Adult *C. furcatus* occur throughout the Gulf of Mexico and Caribbean Sea and to 42 °N in the Gulf Stream (Staiger, 1965). Details of the reproductive periodicity for this species are sketchy. Staiger (1965) suggests that the reproductive period is extended though not necessarily continuous. This is based on the presence of small juveniles in Caribbean waters during February, and from May to July, and another report of a ripe male in November. Eggs are attached to flotsam and *Sargassum* is probably an important reproductive habitat. Juvenile *C. melanurus* and *C. furcatus* were present during April, May and June suggesting that reproduction precedes and continues through these months.

4.8.2 Juvenile Distributions

Cypselurus melanurus has been reported to reach sexual maturity at 25 cm (not specified but assumed to be FL; http://oceanlink.island.net). *Cypselurus furcatus* is a larger fish and probably matures at a greater size. The maximum sizes of fish collected in the dataset place them within

128

the juvenile size ranges and provide an opportunity to assess the distributions of juveniles from April through June.

In January, five individuals of *C. furcatus* were collected at three locations over the outer continental slope in the western and central regions of the study area (Fig. 44). During April, a single juvenile specimen of *C. furcatus* (89 mm FL) was collected well south of the study area at 24.2833 ° N, 84.6667 ° W. During May, this distributional pattern continued with a northward expansion to the shelf-slope break in the western, central and the western half of the northern zone (Fig. 45). One individual was also collected at the south-central boundary of the eastern zone in May (Fig. 45). During June, scattered *C. furcatus* were collected over, or seaward of the outer continental slope in the western, central and eastern zones, and a single fish was taken over the shelf on the north central boundary of the western zone.

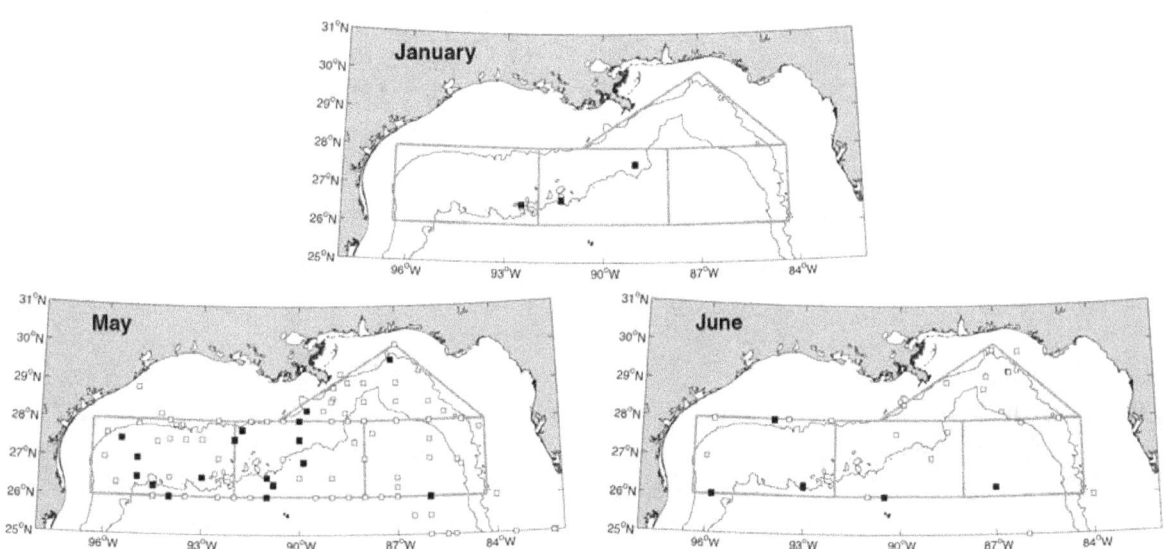

Figure 44. Locations where *Cypselurus furcatus* were present (■) and absent (□) in the study area based on data provided by NOAA (Dr. R.L. Pitman, pers comm.) for the period 1990-1996. Data from January were not part of a systematic cruise survey.

During April, *C. melanurus* were present in the central, eastern and northern regions of the study area (Fig. 44). Within these regions, fish were located in the slope and oceanic waters. By May, *C. melanurus* were distributed throughout the study region and had moved north to the shelf-slope break and on to the shelf (Fig. 44). By June, their distribution pattern appeared to be similar to that observed in May, however, there were fewer stations from which to infer the distributional pattern (Fig. 44).

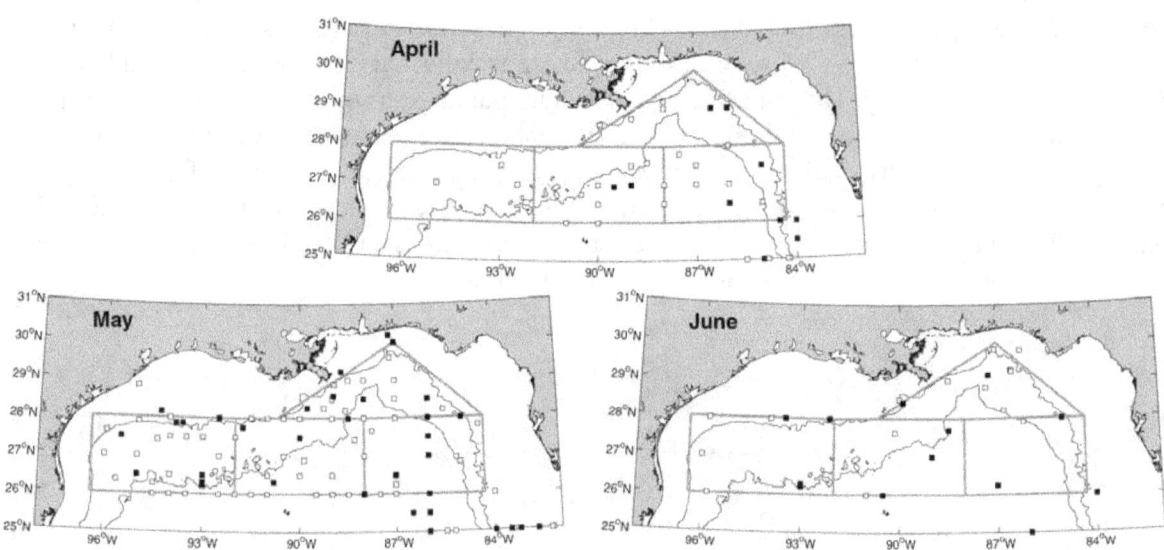

Figure 45. Locations where *Cypselurus melanurus* were present (■) and absent (□) in the study area based on data provided by NOAA (Dr. R.L. Pitman, pers comm.) for the period 1990-1996.

4.8.3 Adult Distributions

There were insufficient data to estimate the distributions of adult flyingfishes of either species. It is likely that they co-occur with the juveniles.

4.8.4 Predicted Juvenile Distributions

Given the potentially greater mobility of juvenile flyingfishes, we applied adult rules for distributions to infer their distributions. Juvenile *C. furcatus* are predicted to be present along the waters over the outer slope in the central and eastern portion of the western zones during January (Fig. 46). Their distribution during January is probably much broader, however, the limited spatial extent of the data prevent further predictions. Since they were present in January and again in May, it is likely that they occur in the study area during the intervening months. By May this species is likely present throughout most of the western and central zones with a sparser distribution in the northern and eastern zones (Fig. 46). In June it is likely found along the southern boundary of the western, central and eastern zones, and possibly inshore as far as the shelf break (Fig. 46).

Scattered *C. melanurus* are predicted to be found in the waters over the outer shelf, slope and oceanic regions of the central, eastern and northern zones during April (Fig. 47). By May and June, this species is likely present throughout most of the study area (Fig. 47).

130

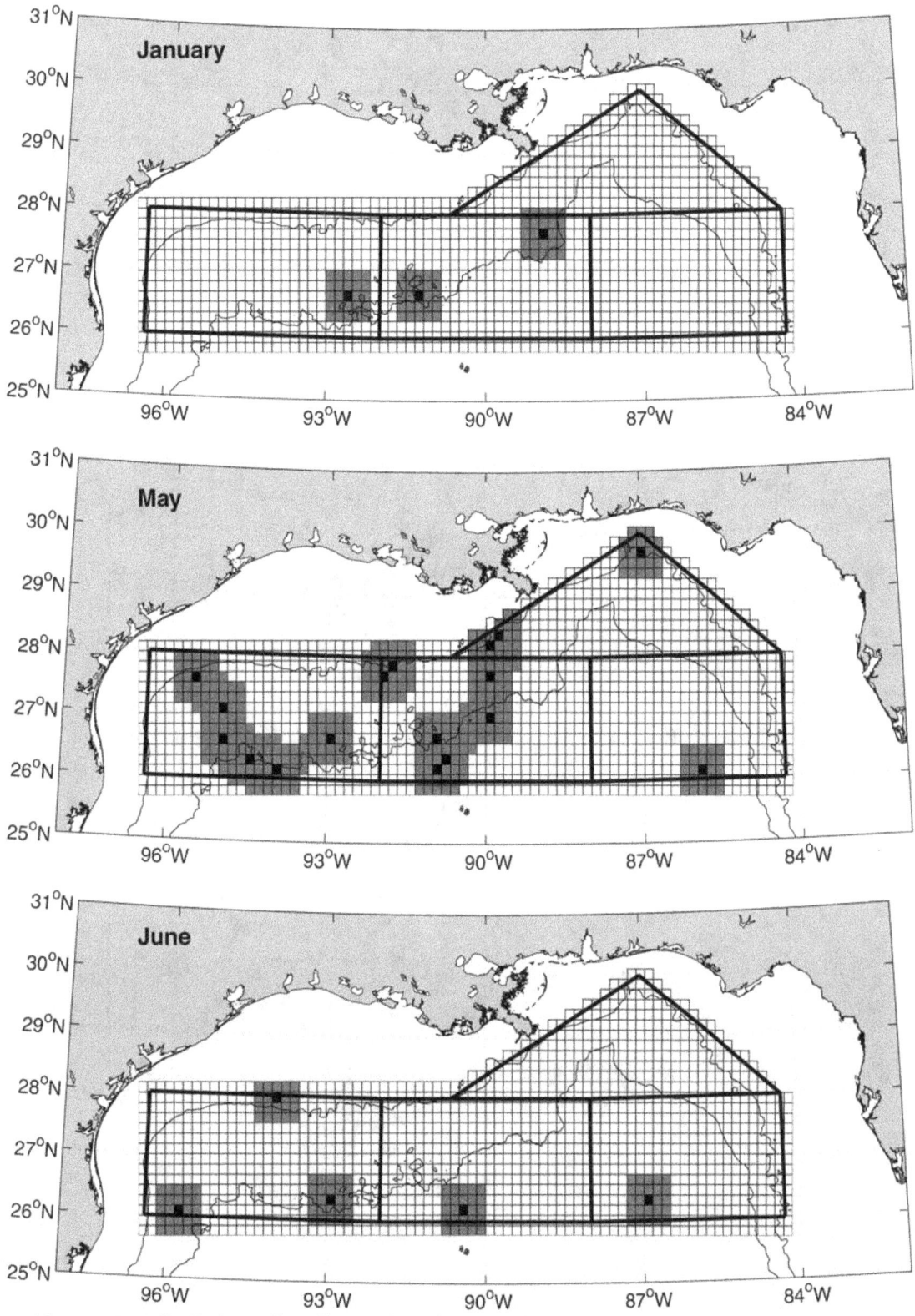

Figure 46. Predicted distributions of juvenile *Cypselurus furcatus* in the study area during January, May, and June. The presence of individuals in each grid cell is indicated as confirmed (■), reasonable inference (▨) or unreported (□).

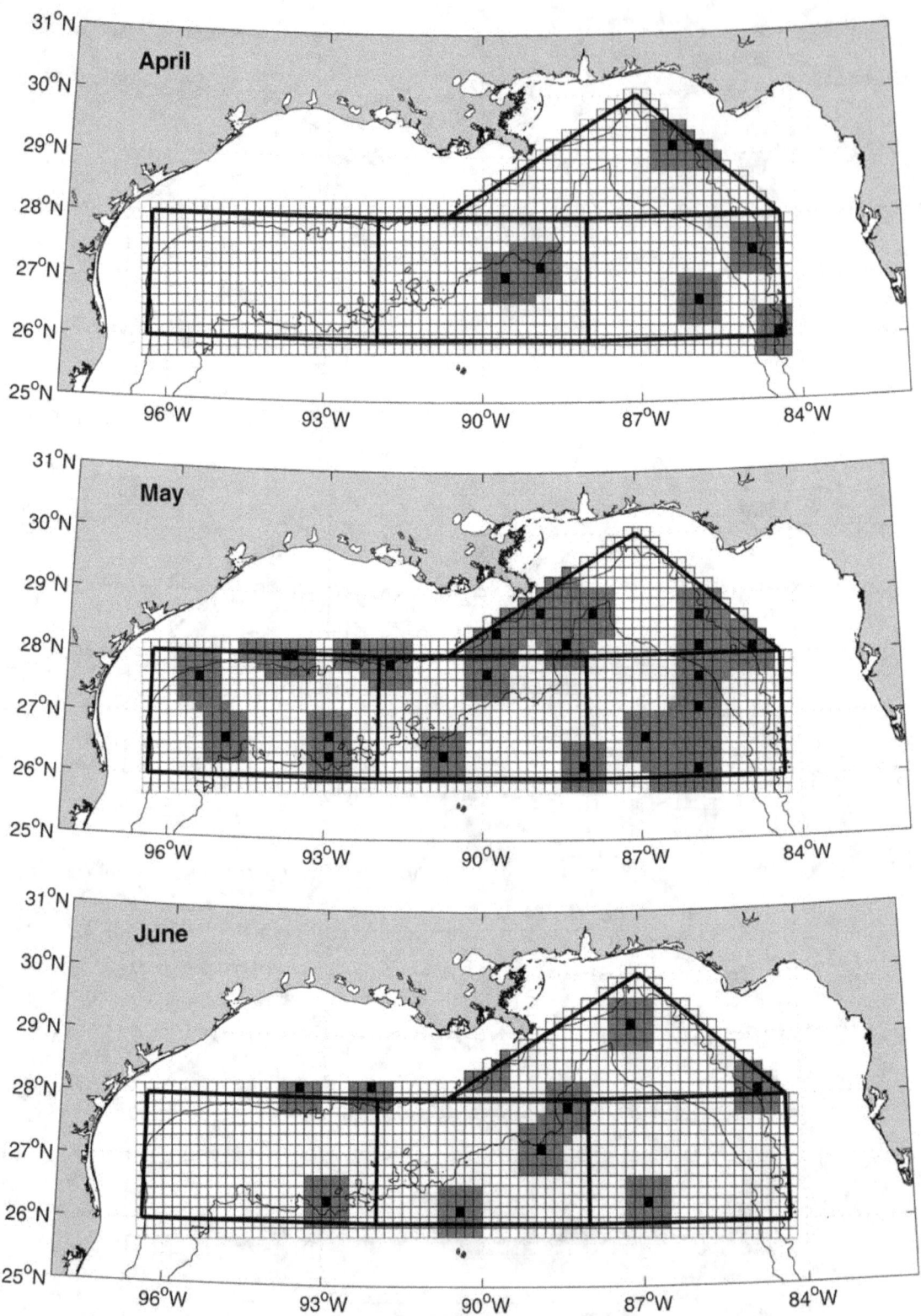

Figure 47. Predicted distributions of juvenile *Cypselurus melanurus* in the study area from April through June. The presence of individuals in each grid cell is indicated as confirmed (■), reasonable inference (▨) or unreported (□).

132

4.9   Ocean Sunfish (*Mola mola*)

Little is known about the ecology of these large pelagic fishes. They have a wide distribution that extends into cool temperate waters in the Atlantic and presumably throughout the pelagic waters of the Gulf of Mexico (Hoese and Moore, 1998). Their diet consists of gelatinous organisms such as jellyfish, ctenophores, and salps. This factor, combined with frequent anecdotal observations of *M. mola* on the ocean surface suggests a distribution that is largely epipelagic, although they have been reported to depths of 1000 m (Block et al., 2000). Ocean sunfish are occasionally present in longline bycatch within the study region (Lee et al., 1994) and made up 0.34% of the commercial pelagic longline catch in the Gulf of Mexico during 1992-1994 (Cramer, 1995). There were insufficient data available to predict the distributions of this taxon in the study area.

4.10   Fishes Associated with the Pelagic *Sargassum* Community

Sargassum or gulfweed is a pelagic complex of two species of brown algae: *S. natans* and *S. fluitans*. Pelagic mats made up of one or both species form floating habitat ranging in size from a few individual plants to rafts hundreds of meters across. These mats propagate vegetatively (Parr 1939) and constitute a stable, buoyant, pelagic community. Physical processes such as Langmuir circulation cells, internal wave packets, and frontal convergences can alter the morphology and size of *Sargassum* mats.

Sargassum mats are permanent or temporary habitat for an estimated 145 species of invertebrates and over 100 species of fish as well as four species of sea turtles and a variety of marine birds. (Atlantic States Marine Fisheries Commission, 1999). Several of these fishes, such as the sargassumfish (*Histrio histrio*), *Sargassum* triggerfish (*Balistes capriscus*), planehead filefish (*Monocanthus hispidus*), chain pipefish (*Syngnathus pelagicus*) and the *Sargassum* seahorse (*Hippocampus ramulosus*) are generally associated only with *Sargassum* communities, while other species such as the ocean sunfish (*Mola mola*) and a variety of carangids are temporary associates (Adams, 1960). Juvenile fishes dominate the community (South Atlantic Marine Fishery Management Council, 1998). Bortone et al. (1977) examined the ichthyofauna of pelagic *Sargassum* collected in the eastern Gulf of Mexico. Numerically, the fish community was dominated by *M. hispidus* (84.5%), *B. capriscus* (6.1%), *H. histrio* (1.8%) and blue runner *Caranx crysos* (1.4%). The ichthyofauna from their most pelagic stations, which were located in water deeper than 183 m included 25 species representing 9 families (Table 2).

Studies in the South Atlantic suggest that the species richness of the *Sargassum* community on the outer continental shelf is greatest during the spring and summer. The diversity in offshore waters was higher than in coastal areas. The spatial distribution of fishes appears to vary around mats with juvenile dolphin, tripletail and flying fishes occupying the periphery, while triggerfish, filefish, and jacks are distributed below the algae, and *Sargassum* fish, pipefish and seahorses are found within the mats (South Atlantic Fishery Management Council, 1998).

Floating *Sargassum* serves as important nursery habitat for larvae and juveniles of a variety of pelagic finfish and foraging habitat for larger juveniles and adult fish. Dolphinfish, wahoos and tunas have all been reported to contain both the *Sargassum* algae and *Sargassum*-associated fauna (Matthews et al. 1977; Manooch and Hogarth, 1983; Manooch and Mason, 1983; Manooch et al. 1984). Recognition of the importance of *Sargassum* to king mackerel, spanish mackerel, cobia and dolphin prompted its designation as essential fish habitat and habitat area of particular concern under the Magnuson-Stevens Act (Atlantic States Marine Fisheries Commission, 1999).

Sargassum is common within the Sargasso Sea, which is bounded by the Gulf Stream, the North Atlantic Current, the Canaries Current and the North Equatorial Current. While *Sargassum* rafts are most common in the Gulf Stream and Florida Current regions, *Sargassum* is also abundant within the Gulf of Mexico (Adams, 1960). Although the biomass of *Sargassum* in the Gulf of Mexico is more patchy than in the Sargasso Sea, densities in the Gulf (up to 1 g m^{-2}) can approach those in the Sargasso Sea (0.2-2 g m^{-2}; Butler et al. 1983). Since *Sargassum* rafts can drift, fragment and coalesce, local prediction of the abundance of *Sargassum* is probably not possible (Parr, 1939). Bortone et al. (1977) speculated that differences in species diversity observed in the Gulf of Mexico relative to the South Atlantic may be a consequence of the sizes of clumps in the two regions. Clumps in the Gulf of Mexico were relatively small compared with those in the South Atlantic.

Studies on the ichthyofauna of *Sargassum* in the south Atlantic (Settle, 1993) indicated a seasonal pattern in fish biomass. While the mean abundance of fishes was highly variable, there were nonetheless, greater abundances of fish during spring relative to fall and winter. Summer abundances were higher than in winter. Settle also found that abundances averaged across seasons, was higher on the inner shelf than the outer shelf.

Table 2 Ichthyofauna Collected from *Sargassum* in Pelagic Waters of the Eastern Gulf of Mexico by Bortone et al. (1977). (All zones were in waters deeper than the 183-m isobath. Zones I–IIIC extended from north to south in their study area (see lower inset figure after Bortone et al. (1977; Fig. 1). After Bortone et al. (1977) Table 1.)

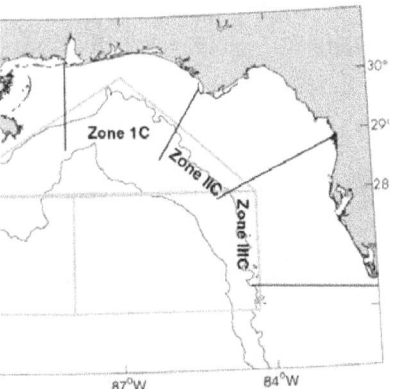

Species	Total	Zone IC	Zone II C	Zone IIIC
Antennariidae				
Histrio histrio	15	2	2	11
Syngnathidae				
Hippocampus erectus	2	-	1	1
Syngnathus louisianae	1	-	1	-
S. pelagicus	12	1	2	9
Carangidae				
Caranx bartholomaei	3	-	1	2
C. crysos	16	2	1	13
C. hippos	1	-	1	-
C. latus	9	9	-	-
Chloroscombrus chrysurus	1	1	-	-
Seriola dumerili	21	3	16	2
S. fasciata	2	-	2	-
S. rivoliana	2	-	-	2
Seriola sp.	1	-	-	1
Coryphaenidae				
Coryphaena hippurus	10	1	5	4
Lobotidae				
Lobotes surinamensis	5	-	-	5
Scombridae				
Scomberomorus cavalla	1	-	-	1
Stromateidae				
Psenes cyanophrys	2	-	-	2
Balistidae				
Aluterus heudeloti	13	4	6	3
Balistes capriscus	158	60	6	92
Canthidermis maculatus	4	3	-	1
Monacanthus ciliatus	1	-	1	-
M. hispidus	585	54	407	124
M. setifer	5	-	-	5
Diodontidae				
Chilomycterus antennatus	1	1	-	-
Diodon holocanthus	2	-	1	1

4.10.1 Sargassumfish (*Histrio histrio*)
4.10.1.1 Adult Distributions
Adult *H. histrio* appear to be obligate residents of *Sargassum* rafts. As is the case with *Sargassum* there is insufficient information to predict their distributions. There is a high probability of their presence wherever *Sargassum* occurs. Given the association of larvae and juveniles with *Sargassum*, the distributions of larvae may be used as an estimate of the locations of adults.

4.10.1.2 Reproduction
Spawning is believed to occur year-round in the Florida Current (Adams, 1960). This is based on the presence of postlarvae in plankton samples during every month except October and that omission was considered a consequence of low sampling effort. Lower water temperatures (possibly water cooler than 18 °C) during winter may interrupt spawning (Adams, 1960). Spawning in the Florida Current occurred from late-August through April (Dooley, 1972).

Fishes larger than 9.0 mm standard length were collected near the surface in close association with *Sargassum* (Adams, 1960) while smaller larvae were reported in samples from as deep as 600 m. These deeper observations may have been a consequence of contaminated samples or non-viable larvae that had been lost from the surface waters while the possibility that early larvae undertook vertical migration could not be evaluated (Adams, 1960).

4.10.1.3 Larval/Juvenile Distributions
Larvae were present throughout the year in the SEAMAP samples with the exception of August (Fig. 48) suggesting that the pattern of year round spawning in the Florida current is also applicable to the Gulf of Mexico. During January larvae were generally present in pelagic waters around the 2000 m isobath with some suggestion of an onshore shift in distributions through March (Fig. 48). In April and May they were again present through middle latitudes of the western, central, and eastern zones. From June through the remainder of the year, larvae were present at scattered locations that were frequently over the shelf. Juveniles settle on to *Sargassum* and they are probably present wherever adults and *Sargassum* are to be found.

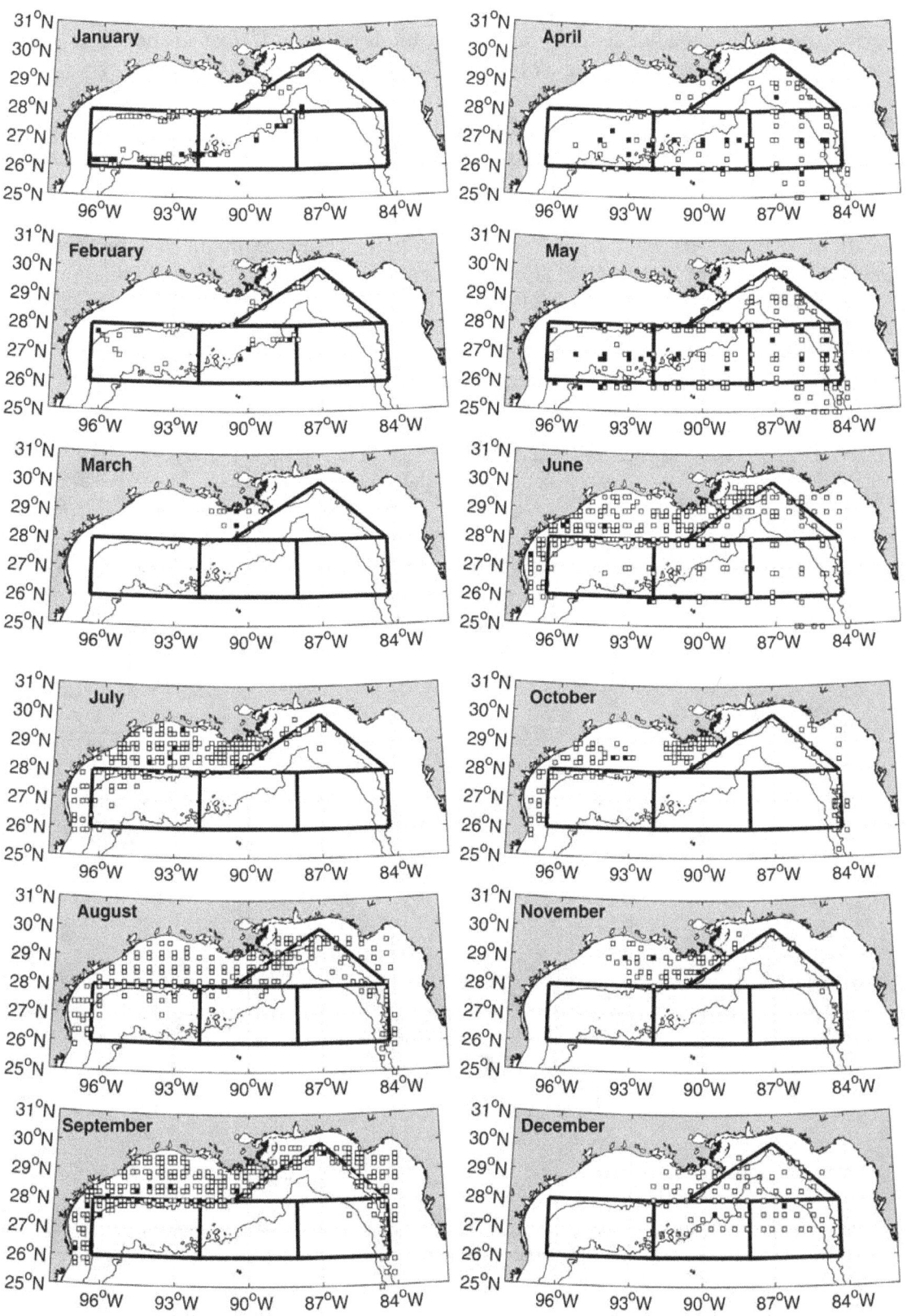

Figure 48. Presence (■) and absence (□) of *Histrio histrio* larvae in the study area from January through December based on SEAMAP ichthyoplankton data.

137

4.10.1.4 Predicted Adult Distributions

In the absence of data on adult distributions, the locations of larval *H. histrio* and any evidence of *Sargassum* provide the best estimate of the probable adult locations (Fig. 48).

4.10.1.5 Predicted Larval/Juvenile Distributions

Larvae are likely present in the study area throughout the year. During January and February, they are presumed to be located offshore in slope and oceanic waters (Fig. 49). This pattern is likely for March as well, although the sampling coverage of the SEAMAP data precludes verification. As the waters warm from April through June, larvae are predicted to be present in all zones (Fig. 49). From July through December, scattered larvae will likely be found in the study area wherever *Sargassum* rafts are located. Juvenile distributions will follow those of adults and *Sargassum*.

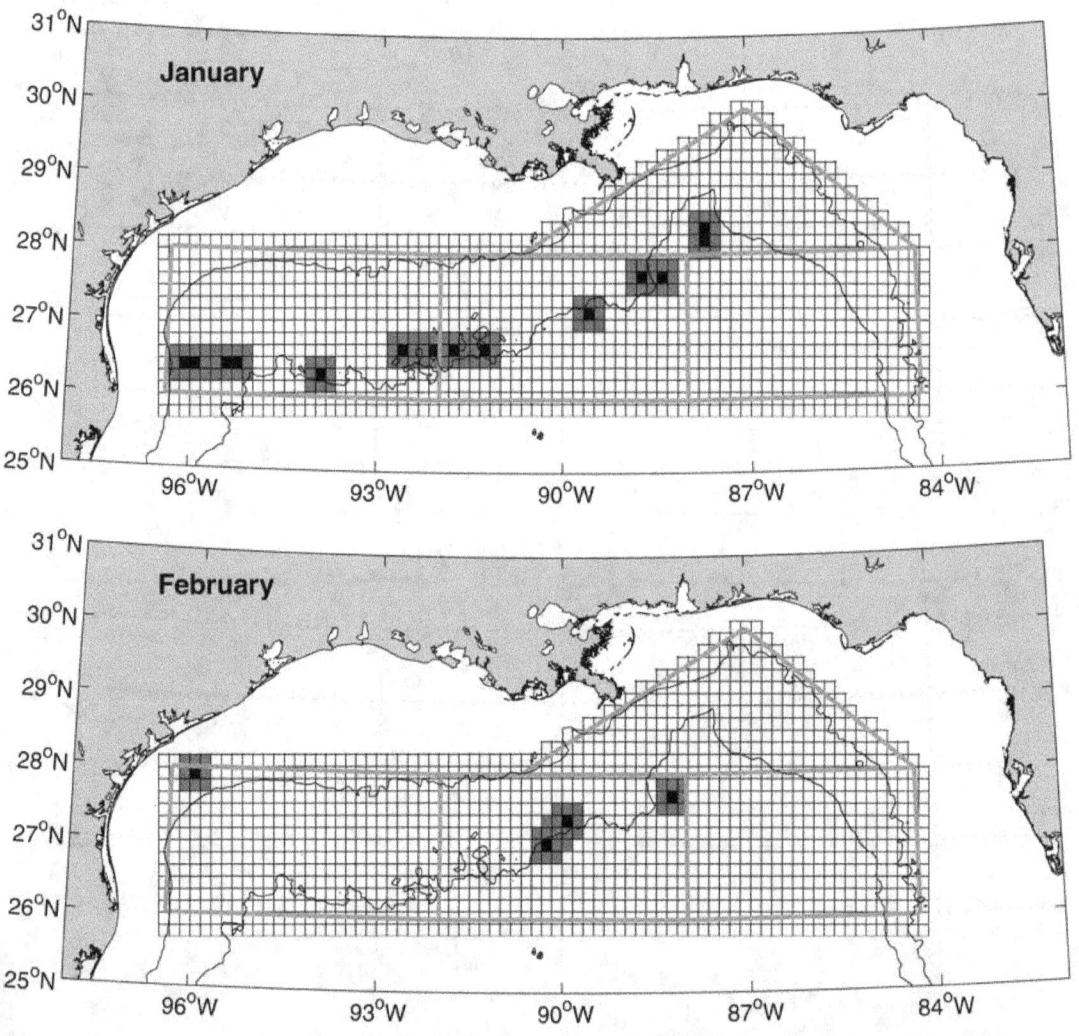

Figure 49.   Predicted distributions of larval *Histrio histrio* in the study area from January through December. The presence of individuals in each grid cell is indicated as confirmed (■), reasonable inference (▨) or unreported (□).

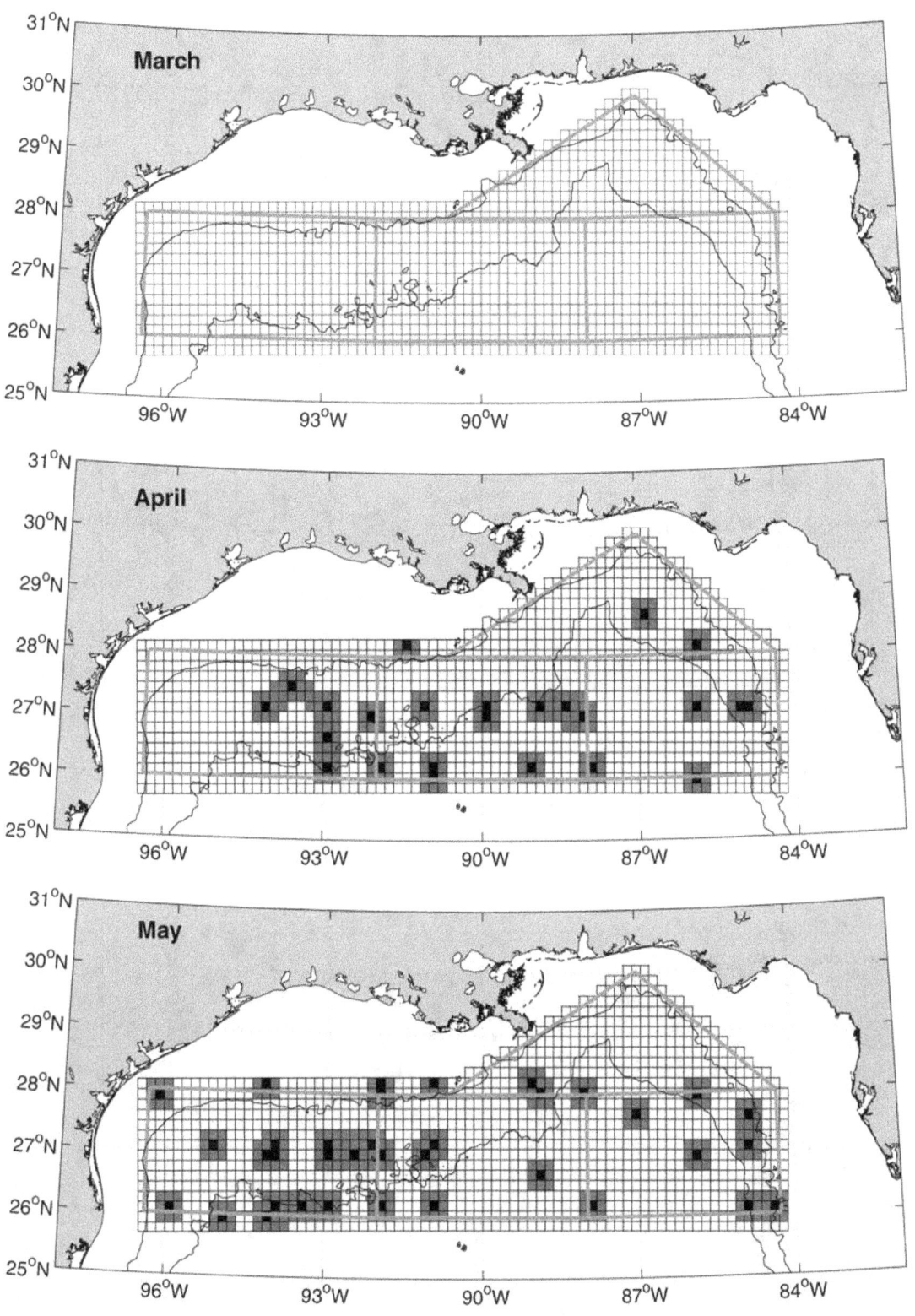

Figure 49. Predicted distributions of larval *Histrio histrio* in the study area from January through December. The presence of individuals in each grid cell is indicated as confirmed (■), reasonable inference (▨) or unreported (□). (continued)

139

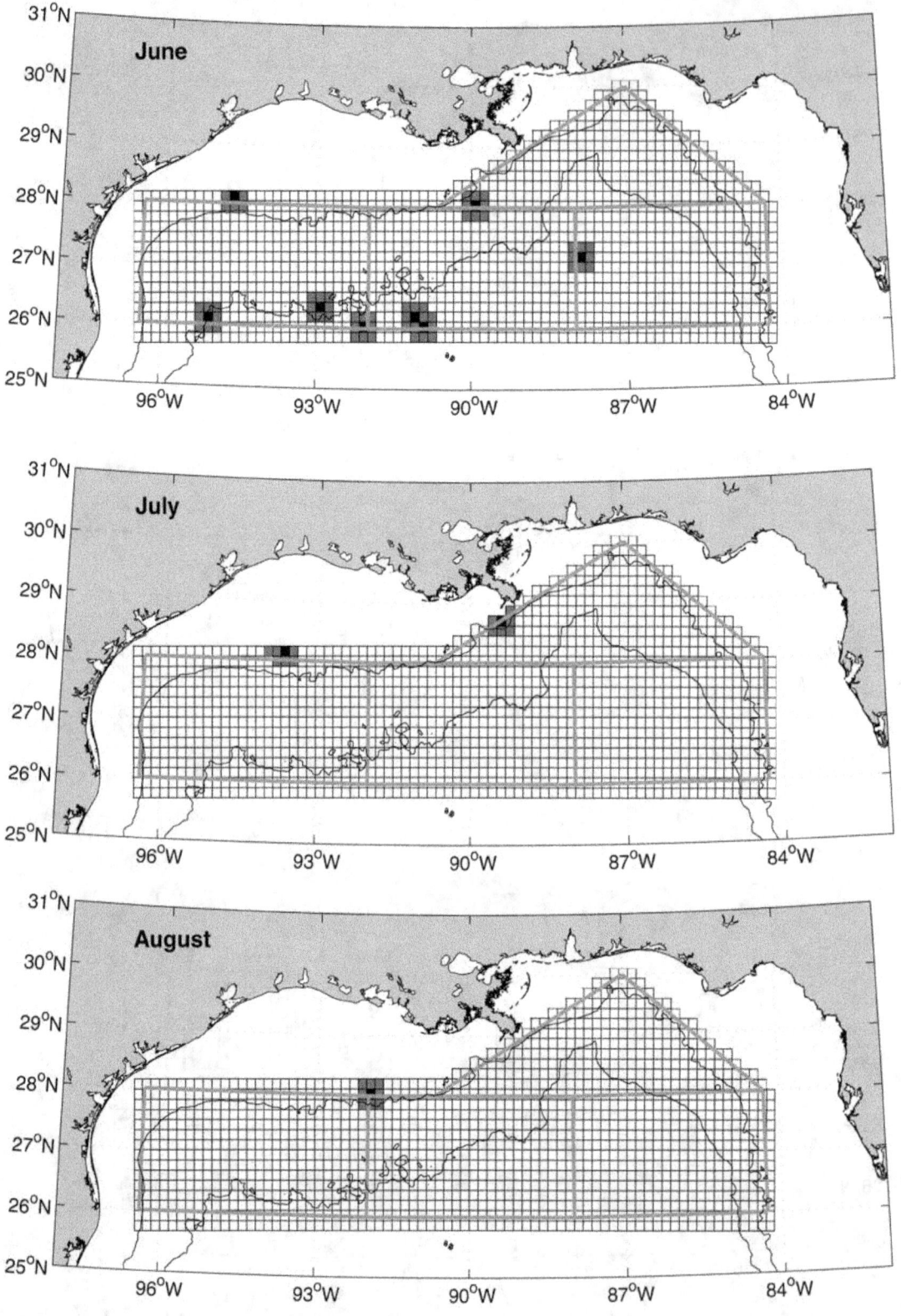

Figure 49. Predicted distributions of larval *Histrio histrio* in the study area from January through December. The presence of individuals in each grid cell is indicated as confirmed (■), reasonable inference (▨) or unreported (□). (continued)

140

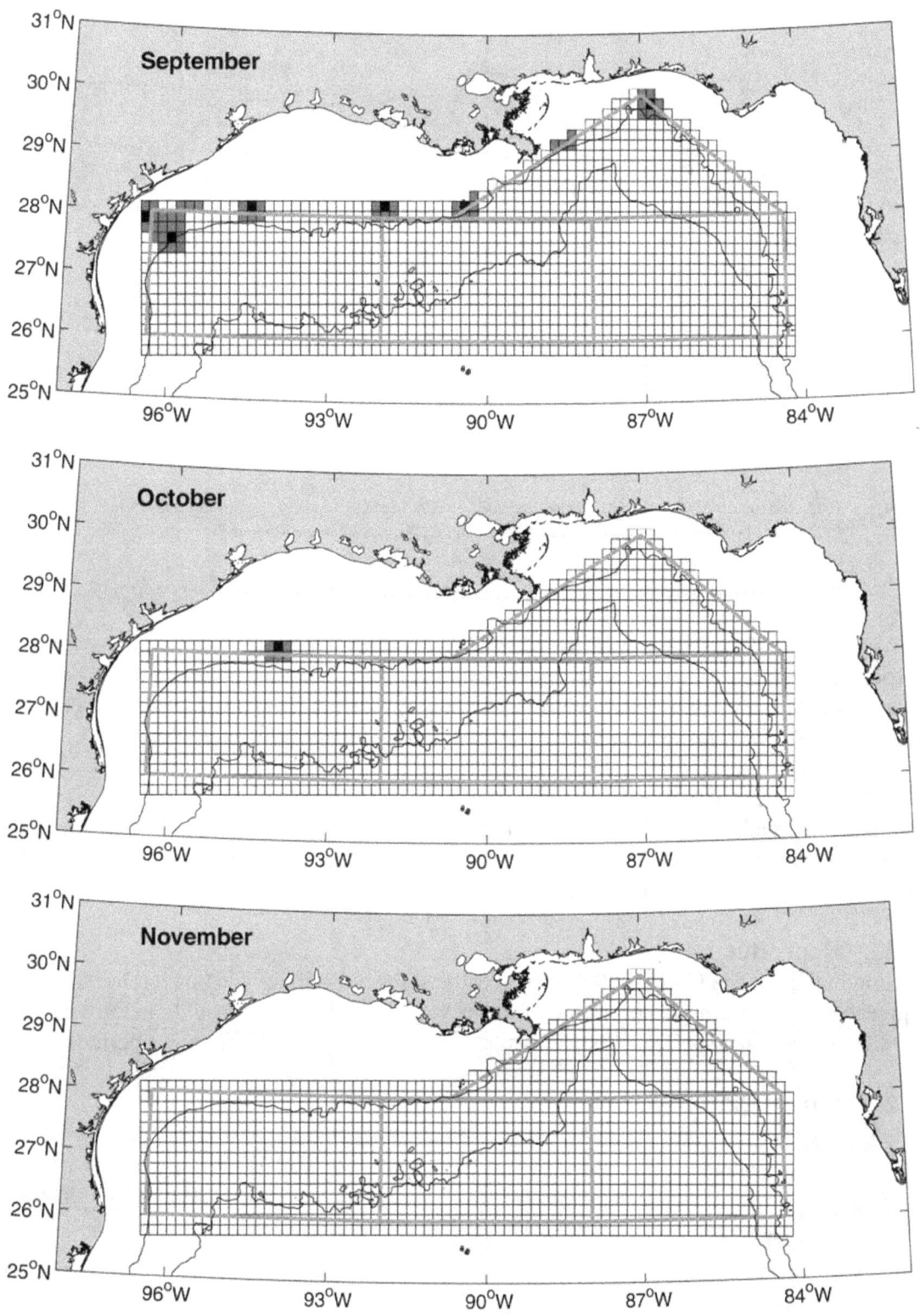

Figure 49. Predicted distributions of larval *Histrio histrio* in the study area from January through December. The presence of individuals in each grid cell is indicated as confirmed (■), reasonable inference (▦) or unreported (□). (continued)

141

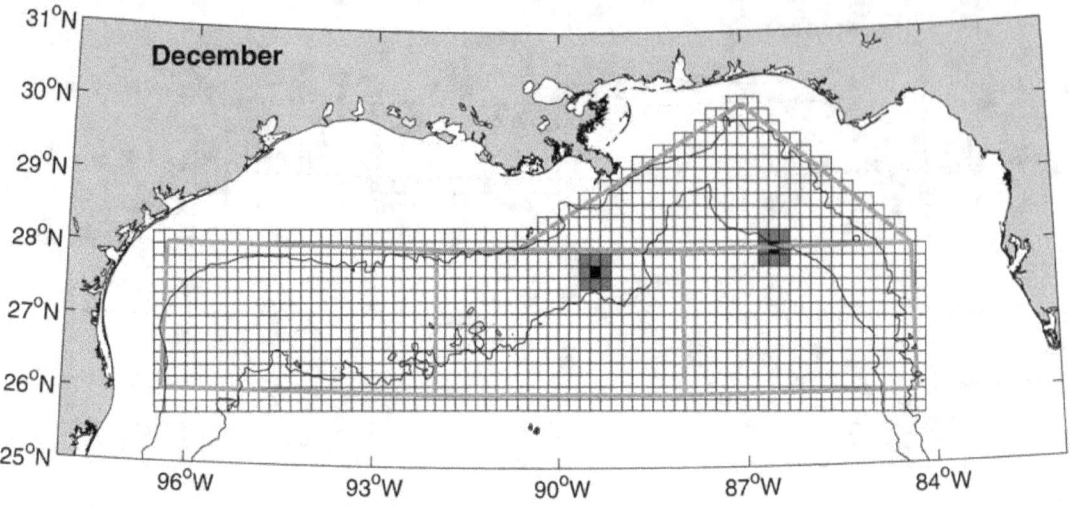

Figure 49. Predicted distributions of larval *Histrio histrio* in the study area from January through December. The presence of individuals in each grid cell is indicated as confirmed (■), reasonable inference (▨) or unreported (□). (continued)

4.10.2   Planehead Filefish (*Monocanthus hispidus*)

This species numerically dominated the ichthyofauna of *Sargassum* patches sampled between May and August in the northeastern Gulf of Mexico by Bortone et al. (1977). The length frequencies of *M. hispidus* ranged from 5-59 mm standard length (mean 20.8 mm).

4.10.2.1   Adult Distributions

The abundance of adults of *M. hispidus* peaked in *Sargassum* in the Florida Current during July and August (Dooley, 1972). As is the case for *H. histrio*, adult *M. hispidus* are likely to be found wherever *Sargassum* rafts are present.

4.10.2.2   Reproduction

Spawning in the Florida Current likely occurred between September and April (Dooley, 1972). The presence of larvae in the northern Gulf from May through June and again in the fall (Ditty et al., 1988) suggests that spawning may occur slightly later with two recruitment periods.

4.10.2.3   Larval Distributions

Larval *M. hispidus* were present in the northern Gulf of Mexico from May through June and again during September and November (Ditty et al., 1988). In SEAMAP samples, larvae were present during June, July and September, however, during July, larvae were collected inshore of the study area (Fig. 50).

142

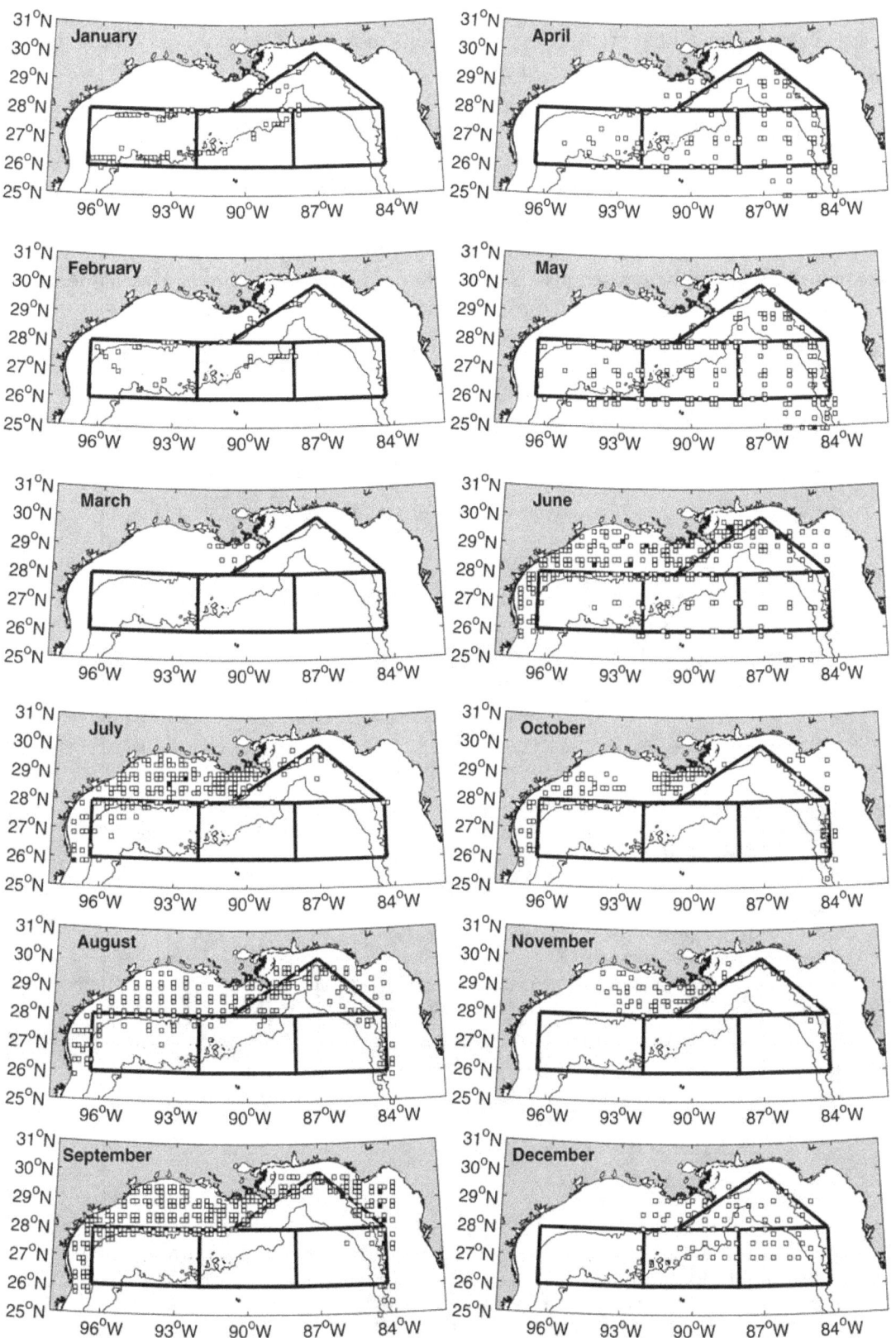

Figure 50. Presence (■) and absence (□) of *Monocanthus hispidus* larvae in the study area from January through December based on SEAMAP ichthyoplankton data.

143

4.10.2.4 Predicted Adult Distributions

There was insufficient data to predict the distributions of adults. They are likely located in the same areas as the larvae and wherever *Sargassum* occurs.

4.10.2.5 Predicted Larval/Juvenile Distributions

The confirmed or reasonably inferred presence of larvae of *M. hispidus* in the study zones was limited to June and September (Fig. 51). Larvae are likely also present during May, July and August because they were detected outside the study area during the two former months, and their presence during September suggests continuity through August. As is the case with *H. histrio*, they are likely to be located near *Sargassum* rafts. The distributions of juveniles will likely co-occur with that of adults and *Sargassum*.

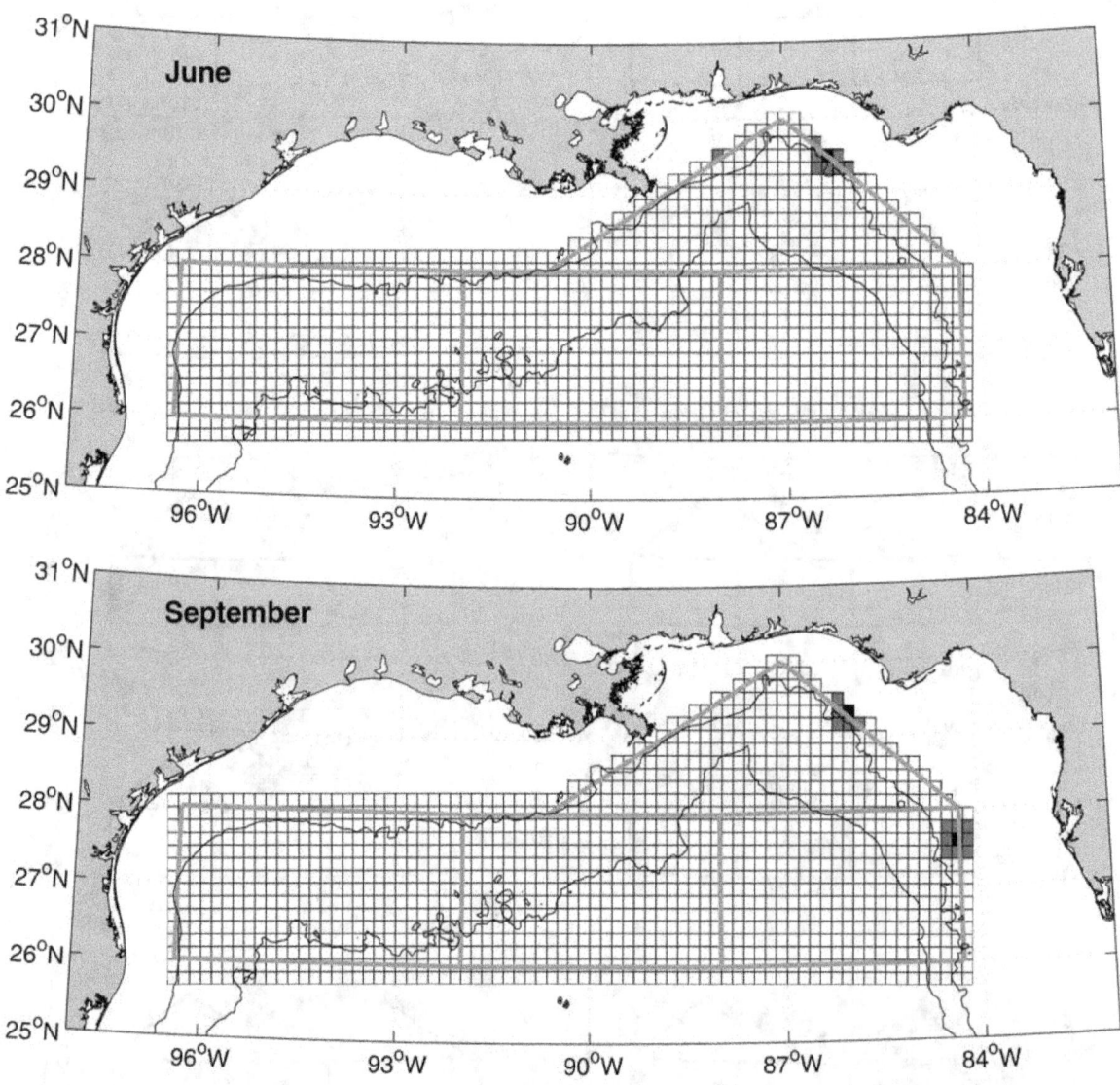

Figure 51. Predicted distributions of larval *Monocanthus hispidus* in the study area during June and September. Larvae were absent from the study zone at other times of the year. The presence of individuals in each grid cell is indicated as confirmed (■), reasonable inference (▨) or unreported (□).

144

4.10.3 Tripletail (*Lobotes surinamensis*)

This species has a broad geographic range that includes the tropical and subtropical regions of the Atlantic, the Caribbean and the Gulf of Mexico (Baughman, 1943). Baughman speculated that the wide distribution of this species is due to the association of young tripletail with *Sargassum* weed. Tripletail appear in Gulf coastal waters from April to early October. Anecdotal records summarized in Baughman (1941) suggest that adult tripletail are common in inshore waters off Mississippi during the summer. Tripletail were less common along the Louisiana coast and it "occurs out in the Gulf, and comes inshore only in the passes at the mouth of the Mississippi River, and in the inlets between the outlying coastal islands, and around these islands." (Baughman, 1941).

4.10.3.1 Adult Distributions

Adult tripletail are found along the Gulf coast from April through early October (Baughman, 1941). They then migrate south during the fall and winter to return in spring (Ditty and Shaw, 1994). Adult tripletail congregate around fixed and floating structures such as sea buoys and navigation beacons. This behavior may result in their aggregation near structure associated with offshore platforms, although this is speculative since there are no records of enhanced densities of tripletail near petroleum platforms.

4.10.3.2 Reproduction

Tripletail reach sexual maturity at 290 and 485 mm total length, for males and females, respectively (Brown-Peterson and Franks, 2001). Tripletails are multiple spawners (Brown-Peterson and Franks, 2001). Baughman (1941) suggested that based on the presence of ripe ovaries in three specimens, spawning in the northern Gulf takes place from June-August. This spawning periodicity was confirmed by Brown-Peterson and Franks (2001) who reported peak spawning during July. Spawning likely occurs offshore since all records of larvae smaller than 5 mm reviewed by Ditty and Shaw (1994) were from stations over the outer shelf and in oceanic waters.

4.10.3.3 Larval/Juvenile Distributions

Larvae were present almost exclusively from July through September from surface waters of the northern Gulf (Ditty and Shaw, 1994). Over 75% of the larvae in distribution records reviewed by Ditty et al. (1994) were collected from waters ≥ 28.8 °C and salinities ≥ 30.3 psu. The SEAMAP dataset indicated that larvae were present from May to October with evidence of a northward shift in their distribution as the waters of the northern Gulf warm during summer (Fig. 52). Juvenile tripletail become associated with floating seaweed and other flotsam at a size of between 10-20 mm TL (Uchida et al., 1958). Juveniles were present in *Sargassum* from the Florida Current region during summer through winter with a peak during early fall (Dooley, 1972). Laboratory studies of the growth rates of captive early-juvenile tripletail ranging from 45-115 mm TL indicated a mean growth rate of 1.4 mm d^{-1} (Franks et al., 2001). These studies began on July 31 and extended for up to 210 d during which time, the water temperatures ranged from 25.2-29 °C. Their samples were collected from a site 4 km south of West Ship Island, Mississippi.

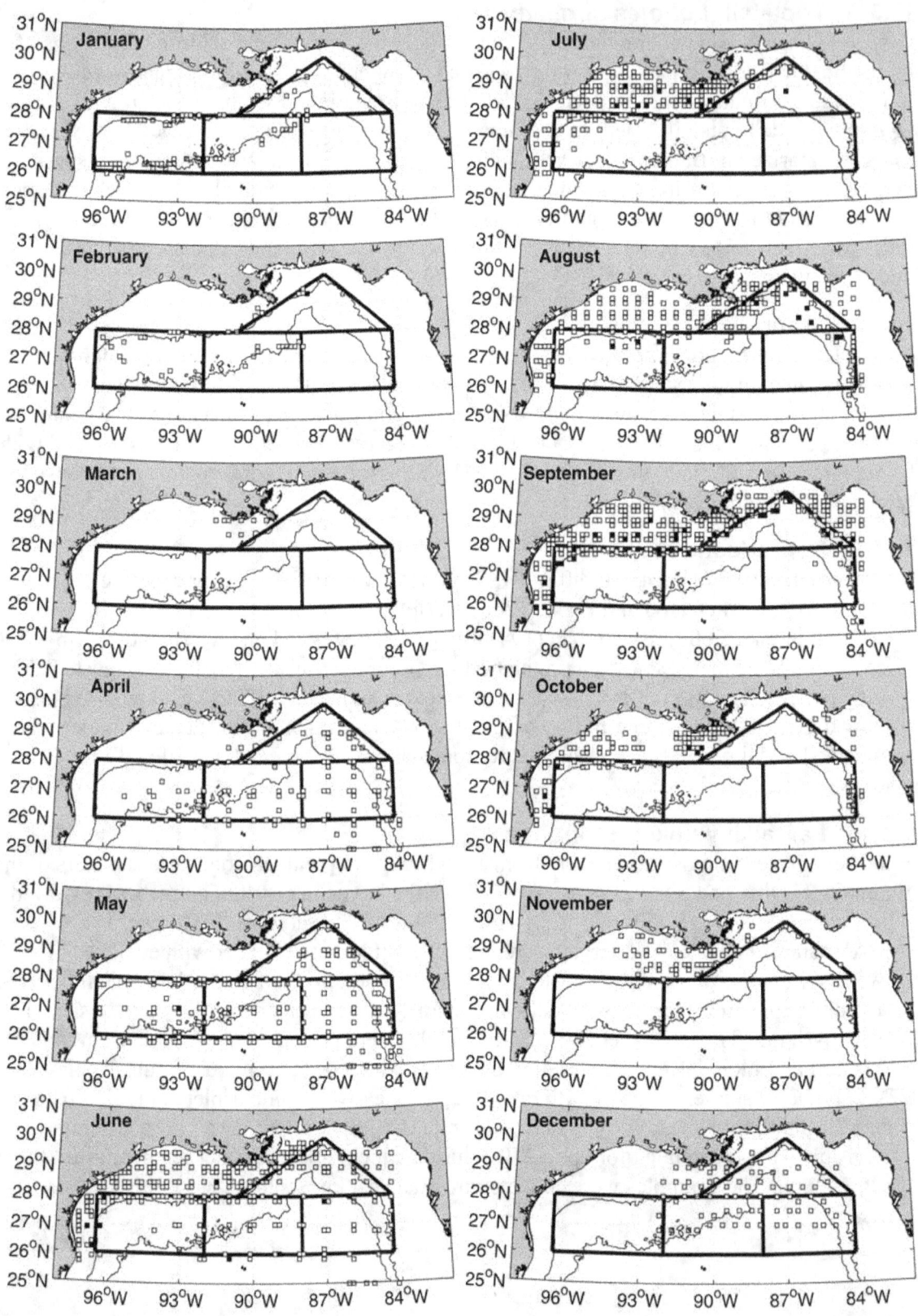

Figure 52. Presence (■) and absence (□) of *Lobotes surinamensis* larvae in the study area from January through December from SEAMAP ichthyoplankton data.

146

4.10.3.4 Predicted Adult Distributions

There were insufficient data to predict the distributions of adult tripletail in the study area.

4.10.3.5 Predicted Larval/Juvenile Distributions

Larval tripletail are predicted to be present over the shelf during February and in the slope water during May (Fig. 53). Their frequency of occurrence increases over the summer when they are present in scattered locations during June and July. By August they will primarily be found in the northern zone and common along the 200 m isobath by September (Fig. 53). Juvenile distributions may parallel those of larvae.

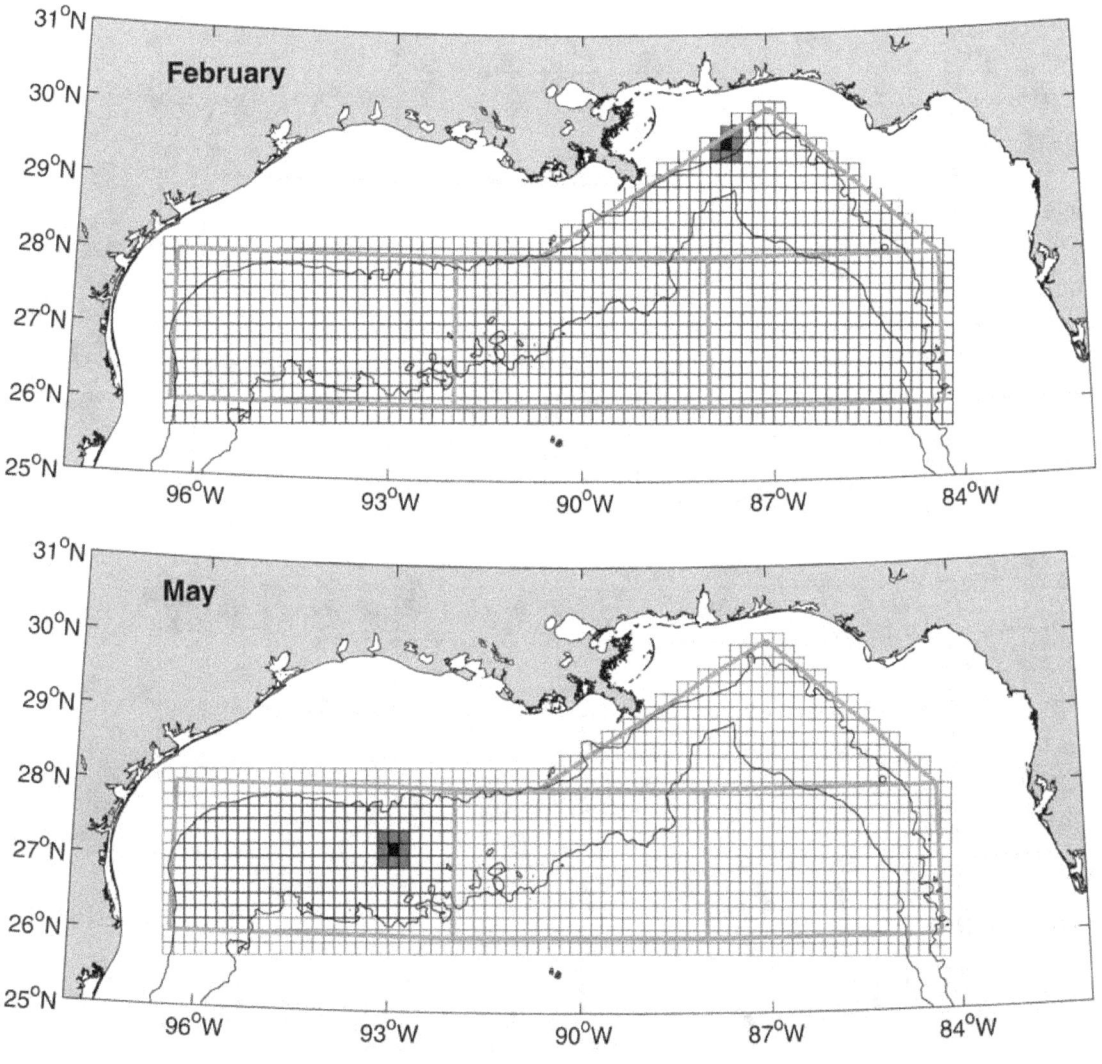

Figure 53. Predicted distributions of larval *Lobotes surinamensis* in the study area during February, May, June, July, August, and September. Larvae were absent from the study zone during March and April. The presence of individuals in each grid cell is indicated as confirmed (■), reasonable inference (▨) or unreported (□).

147

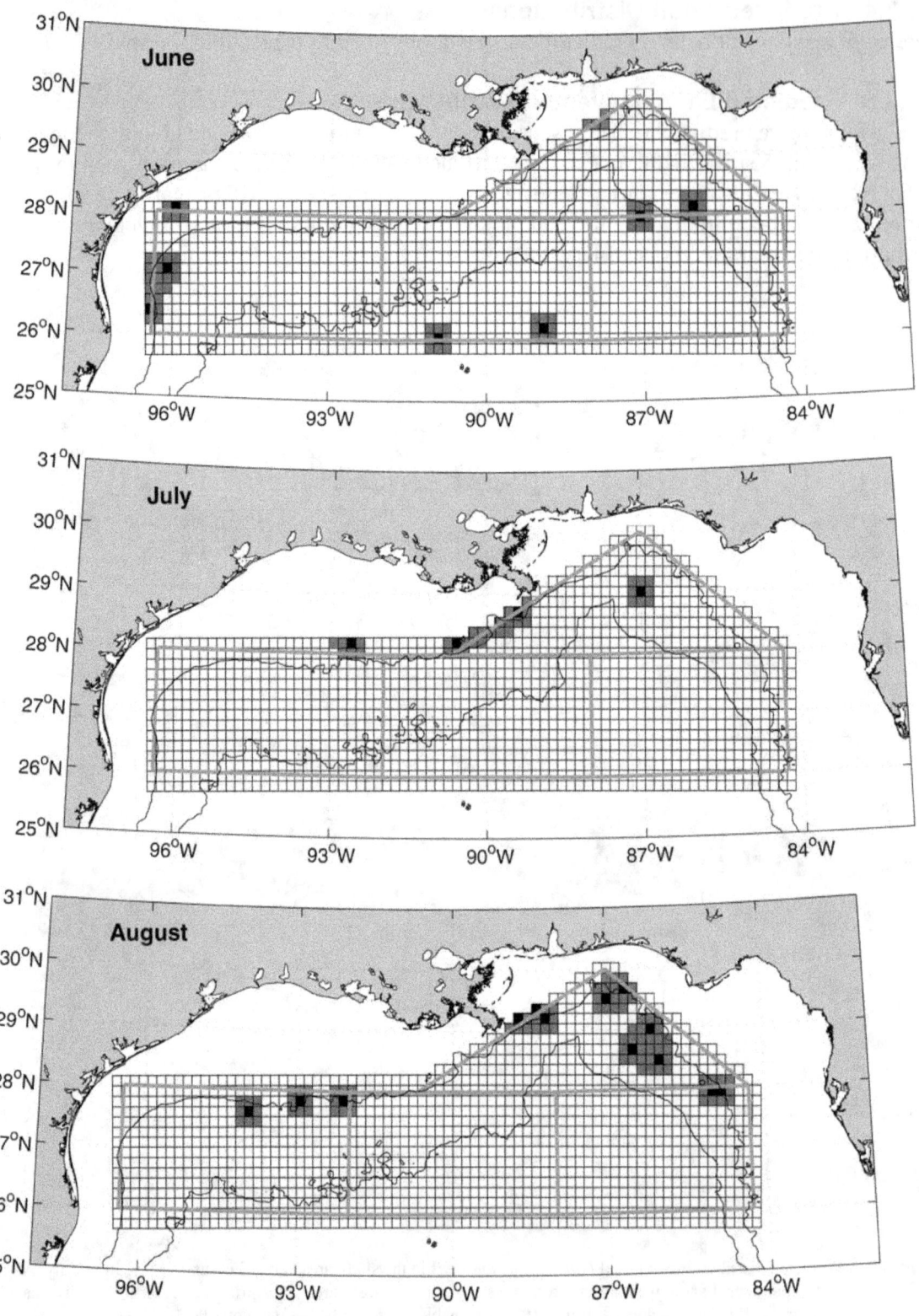

Figure 53. Predicted distributions of larval *Lobotes surinamensis* in the study area during February, May, June, July, August, and September. Larvae were absent from the study zone during March and April. The presence of individuals in each grid cell is indicated as confirmed (■), reasonable inference (▓) or unreported (□). (continued)

148

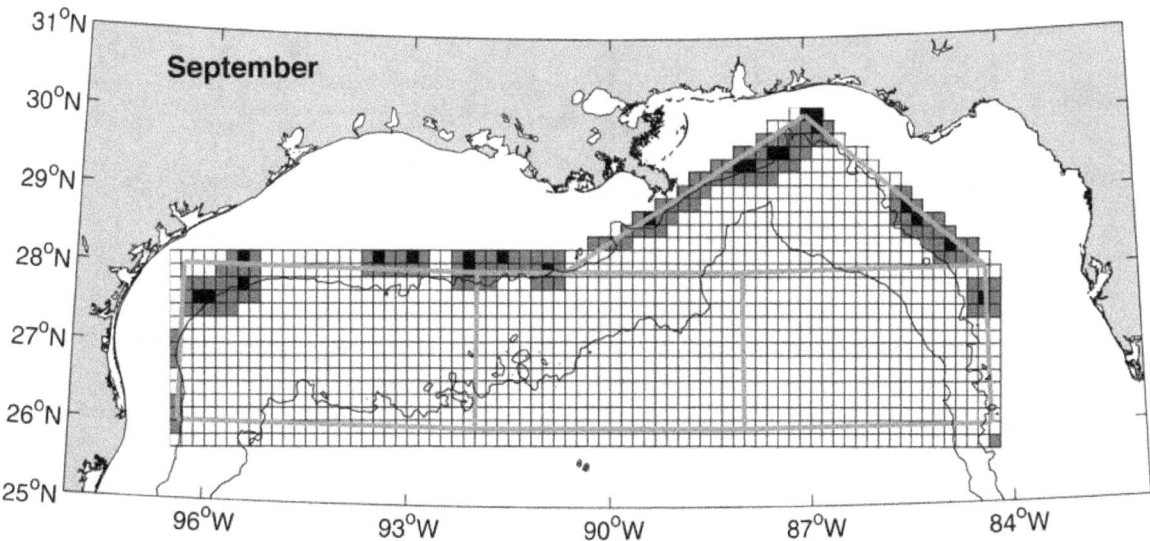

Figure 53. Predicted distributions of larval *Lobotes surinamensis* in the study area during February, May, June, July, August, and September. Larvae were absent from the study zone during March and April. The presence of individuals in each grid cell is indicated as confirmed (■), reasonable inference (▨) or unreported (□). (continued)

149

5   Utilization of the Data

These data provide an estimate of where different life-history stages of selected species are likely to be found on a month-by-month basis. We have subdivided our study area into a grid containing 10 minute x 10 minute cells (~10 nmile x 10 nmile). This grid was then classified into confirmed, reasonable inference and not-reported categories based on the datasets and publications examined. It will be obvious to the reader that it is impossible to state with any certainty whether a particular individual of a given stage and species will actually be present in a confirmed or reasonable inference cell in a particular month. Nor is it possible to estimate the numerical densities that are likely to be present in each of the cells. The distributions of these organisms are likely to be highly patchy in time and space and will exhibit patchiness on scales smaller than the grid cell sizes. Spatial distributions of larval scombrids off Puerto Rico and the Virgin Islands illustrate how patchy densities of these organisms can be on small spatial scales (Hare et al. 2001). The available data on the vertical distributions of the targets species are too limited to make definitive predictions of their location in the water column. In general, larvae and juveniles will be located near the surface where they are vulnerable to petroleum spills. The adults are also located in the pelagic waters and frequently near the surface, however, their mobility and capability to forage in deeper waters makes them far less vulnerable to surface spill toxicity.

With these caveats in mind, the dataset can be utilized to predict whether a given stage (larvae/juveniles or adults) of our target species is likely to be present in the area of a spill. The application of the data in this regard is straightforward. Data for each taxon and stage is provided in a Microsoft Excel spreadsheet separate worksheets for each taxon and life history state. A series of files are provided for each month of the year. There is one set for adults and a second set for larvae/juveniles. Within each file there are worksheets for each species containing latitude and longitude coordinates (Fig. 54) and a matrix of cells containing distributional data are encoded with values of 2 (confirmed), 1 (reasonable inference), or 0 (unreported) (Fig. 54). The cells corresponding to the study area are colored yellow. Within each monthly Excel spreadsheet for either larvae/juveniles or adults there is a worksheet called Distribution Calculator that provides a means of querying the data (Fig. 55). Individual worksheets for each taxon are accessible via tabs at the bottom of the spreadsheet page (Figs. 54, 55).The Distribution Calculator relies on two additional worksheets called Internal 1 and 2 that should not be altered or moved. The Distribution Calculator worksheet allows the user to enter the western and southern coordinates of the location to be queried (Fig. 55). The calculator will query each taxon for the month in question and return the predicted distribution of adults or larvae/juveniles of each taxon at the location of interest.

151

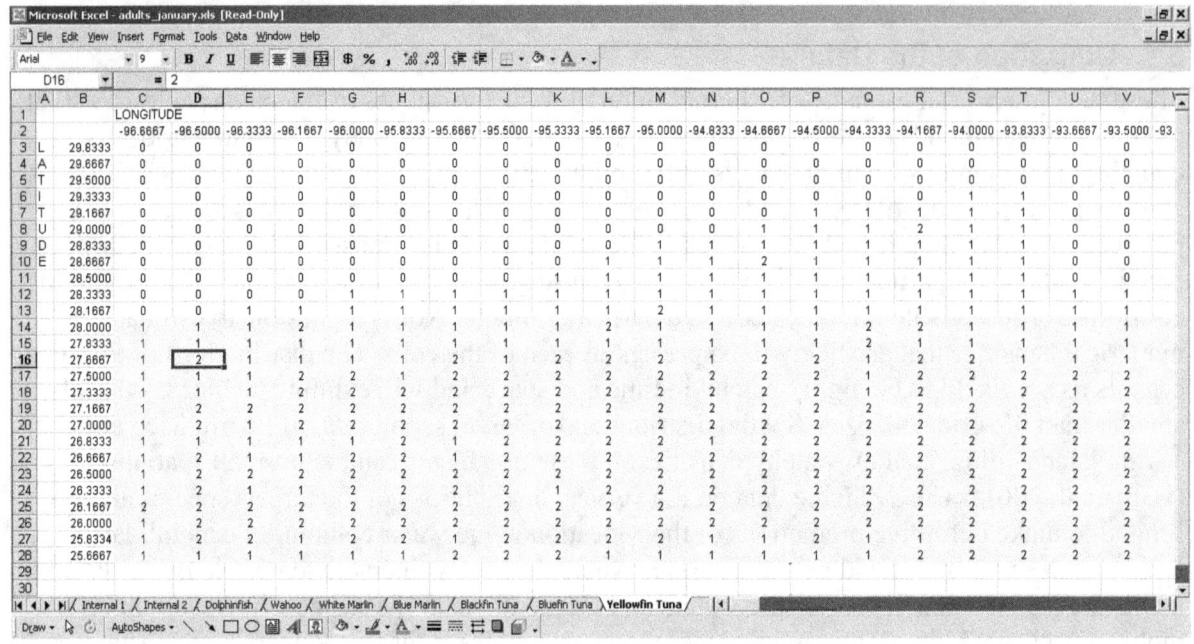

Figure 54. A portion of a Microsoft Excel spreadsheet containing distributional data (confirmed presence, reasonable inference or unreported) for adult Yellowfin Tuna in January. The section of the spreadsheet shaded in pale yellow is the region within the study area. The left column contains latitude coordinates indicating the southern boundary of the corresponding cells in the distributional matrix. The top row contains longitude coordinates indicating the western edge of each 10 x 10 nautical mile cell.

	A	B	C	D	E
1					
2				Example	
3	Enter the spill latitude using decimal degrees:	29		27.3	
4	Enter the spill longitude using decimal degrees	-88		-88.4	
5					
6					
7	**Adults**		**Cell Code**	**Classification**	
8	*Thunnus albacares* (Yellowfin Tuna)		2	Confirmed	
9	*Thunnus thynnus* (Bluefin Tuna)		1	Reasonable Inference	
10	*Thunnus atlanticus* (Blackfin Tuna)		2	Confirmed	
11	*Makaira nigricans* (Blue Marlin)		2	Confirmed	
12	*Tetrapturus albidus* (White Marlin)		1	Reasonable Inference	
13	*Acanthocybium solanderi* (Wahoo)		2	Confirmed	
14	*Coryphaena hippurus* (Dolphinfish)		2	Confirmed	
15					

Figure 55. The Distribution Calculator worksheet within a Microsoft Excel spreadsheet for adult fishes from June. By entering latitude and longitude coordinates, the calculator returns the probable distribution of each taxon in the location.

152

6   References

Adams, C. 1996. An overview of the commercial and recreational fisheries industries within the Gulf of Mexico. The Southern Business and Economic Journal 19: 246-260.

Adams, J. 1960. A contribution to the biology and postlarval development of the *Sargassum* fish, *Histrio histrio* (Linaeus), with a discussion of the *Sargassum* complex. Bulletin of Marine Science of the Gulf and Caribbean 10: 55-82.

Atlantic States Marine Fisheries Commission. 1999. Fishery management council prohibits *Sargassum* seaweed harvest. Habitat Hotline Atlantic 6: 1-2.

Baglin Jr., R.E. 1982. Reproductive biology of western Atlantic bluefin tuna. Fishery Bulletin 80: 121-1334.

Barry A. Vittor & Associates, Inc. 1985. Tuscaloosa Trend Regional Data Search and Synthesis Study (Volume I-Synthesis Report). Final Report Submitted to Minerals Management Service, Metairie, LA. Contract No. 14-12-001-30048. xxvi + 477 p.

Baughman, J. 1941. On the occurrence in the Gulf coast waters of the United States of the tripletail *Lobotes surinamensis*, with notes on its natural history. The American Naturalist 75: 569-579.

Baughman, J. 1943. Additional notes on the occurrence and natural history of the tripletail, *Lobotes surinamensis*. American Midland Naturalist 29: 365-370.

Beardsley Jr., G. J. 1967. Age, growth, and reproduction of the dolphin, *Coryphaena hippurus*, in the straits of Florida. Copeia 441-451.

Beardsley, G. and R. Conser. 1981. An analysis of catch and effort data from the U.S. recreational fishery for billfishes (Istiophoridae) in the Western North Atlantic Ocean and Gulf of Mexico, 1971-78. Fishery Bulletin 79: 49-68.

Block, B., D. Booth and F. Carey. 1992. Depth and temperature of the blue marlin, *Makaira nigricans*, observed by acoustic telemetry. Marine Biology 114: 175-183.

Block, B.A., D.P. Costa, G. Boehlert, and R. Kochevar. 2000. Final report. A report on the tagging of Pacific pelagics (TOPP) workshop. A pilot project for the census of marine life. 81p.

Bortone, S.A., P.A. Hastings and S.B. Collard. 1977. The pelagic-*Sargassum* ichthyofauna of the eastern Gulf of Mexico. Northeast Gulf Science 1: 60-67.

Brown-Peterson, N.J., J.S. Franks, and A.M. Burke. 2000. Preliminary observations on the reproductive biology of wahoo, *Acanthocybium solandri*, from the northern Gulf of Mexico and Bimini, Bahamas. Proceedings of the Gulf and Caribbean Fisheries Institute 51: 414-427.

Brown-Peterson, N.J. and J.S. Franks. 2001. Aspects of the reproductive biology of tripletail, *Lobotes surinamensis*, in the northern Gulf of Mexico. Proceedings of the Gulf and Caribbean Fisheries Institute 52: 586-597.

Brusher, H.A. and B.J. Palko. 1987. Results from the 1984 and 1985 charterboat surveys in southeastern U.S. waters and the U. S. Caribbean Sea. Marine Fisheries Review 49: 109-117.

Butler, J., B. Morris, J. Cadwaller and A. Stoner. 1983. Studies of *Sargassum* and the *Sargassum* community. Bermuda Biological Station Special Publication 22: 1-85.

Christmas, J.Y., A. Perry and R.S. Waller. 1974. Investigations of coastal pelagic fishes completion report. National Marine Fisheries Service, 105p.

Clay, D. 1991. Atlantic bluefin tuna (*Thunnus thynnus thynnus* (L.)): a review. In, R.B. Deriso and W. H. Bayliff (eds.) World meeting on stock assessment of bluefin tunas: strengths and weaknesses, pp. 91-179. Inter-American Tropical Tuna Commission Special Report No. 7, 357 pp.

Collete, B.B. and C.E. Nauen. 1983. FAO Species Catalogue Volume 2: Scombrids of the World. United Nations Food and Agriculture Organization, Rome.

Cramer, J. 1995. Large Pelagic Logbook Newsletter – 1994. NOAA Technical Memorandum NMFS-SEFSC-378.

Cramer, J. and G.P. Scott. 1997. Standardized catch rates for large bluefin tuna, *Thunnus thynnus*, from the U.S. pelagic longline fishery in the Gulf of Mexico and off the Florida east coast. Collective Volume of Scientific Papers- International Commission for the Conservation of Atlantic Tuna 46: 246-251.

Ditty, J.G, G.G. Zieske and R.F. Shaw. 1988. Seasonality and depth distribution of larval fishes in the northern Gulf of Mexico above latitude 26 degrees N. Fishery Bulletin 86: 811-823.

Ditty, J. G. and R.F. Shaw. 1994. Larval development of tripletail, *Lobotes surinamensis* (Pisces: Lobotidae), and their spatial and temporal distribution in the northern Gulf of Mexico. Fishery Bulletin 92: 33-45.

Ditty, J.G., R.F. Shaw, C.B. Grimes and J.S. Cope. 1994. Larval development, distribution and abundance of common dolphin and pompano dolphin in the northern Gulf of Mexico. Fishery Bulletin 92: 275-291.

Dooley, J.K. 1972. Fishes associated with the pelagic *Sargassum* complex, with a discussion of the *Sargassum* community. Contributions in Marine Science 16: 1-32.

Dugas, R.J., V. Guillory and M. Fischer. 1979. Oil rigs and offshore sport fishing in Louisiana. Fisheries 4: 2-10.

Fisher, W., Ed. 1978. FAO Species Identification Sheets for Fishery Purposes Western Central Atlantic (Fishing Area 31). Rome, Food and Agriculture Organization of the United Nations.

Franks, J.S., J.T. Ogle, J.R. Hendon, D.N. Barnes and L.C. Nicholson. 2001. Growth of captive juvenile tripletail *Lobotes surinamensis*. Gulf and Caribbean Research 13: 75-78.

Gibbs, R. H. and B. B. Collette. 1959. On the identification, distribution, and biology of the dolphins, *Coryphaena hippurus* and *C. equiselis*. Bulletin of Marine Science of the Gulf and Caribbean 9: 117-152.

Goldberg, S. and H. Herring-Dyal. 1981. Histological gonad analyses of late summer-early winter collections of bigeye tuna, *Thunnus obesus*, and yellowfin tuna, *Thunnus albacares*, from the Northwest Atlantic and the Gulf of Mexico. NOAA, NOAA Tech. Memo. NOAA-TM-NMFS-SWFC-14: 9p.

Goodwin IV, J.M. and J.H. Finucane. 1985. Reproductive biology of the blue runner (*Caranx crysos*) from the eastern Gulf of Mexico. Northeast Gulf Science 7: 139-146.

Grimes, C. and K. Lang. 1992. Distribution, abundance, growth, mortality and spawning dates of yellowfin tuna, *Thunnus albacares*, larvae around the Mississippi River discharge plume.

Collective Volume of Scientific Papers- International Commission for the Conservation of Atlantic Tuna 38: 177-194.

Hare, J.A., D.E. Hoss, A.B. Powell, M. Konieczna, D.S. Peters, S.R. Cummings, and R. Robbins. 2001. Larval distribution and abundance of the family Scombridae and Scombrolabracidae in the vicinity of Puerto Rico and the Virgin Islands. Bulletin of the Sea Fisheries Institute, Gydnia, 153: 13-29.

Hassler, W.W. and R.P. Rainville. 1975. Techniques for hatching and rearing dolphin, *Coryphaena hippurus*, through larvae and juvenile stages. University of North Carolina Sea Grant Program Publication UNC-SG 75-31, 17 p.

Hoese, H. and R. Moore 1998. Fishes of the Gulf of Mexico: Texas, Louisiana, and adjacent waters. College Station, Texas A&M University Press.

Honma, M., T. Matsumoto and H. Kono. 1985. Comparison of two abundance indices based on Japanese catch and effort data by one-degree and five-degree squares for Atlantic bluefin tuna in the Gulf of Mexico. Collective Volume of Scientific Papers- International Commission for the Conservation of Atlantic Tuna 22: 254-264.

Holland, K.N., R.W. Brill, and R.K.C. Chang. 1990. Horizontal and vertical movements of Pacific blue marlin captured and released using sportfishing gear. Fishery Bulletin, 88: 397–402.

Keenan, S.F., M.C. Benfield, and R.F. Shaw. 2003. Zooplanktivory by blue runner *Caranx crysos*: an energetic subsidy to Gulf of Mexico fish populations at petroleum platforms. American Fisheries Society Symposium, 36:167-180.

Kelley, S., J.V. Gartner Jr., W.J. Richards,and L. Ejsymont. 1990. SEAMAP 1986 – Ichthyoplankton. Larval Distribution and Abundance of Engraulididae, Carangidae, Clupeidae, Gobiidae, Lutjanidae, Serranidae, Coryphaenidae, Istiophoridae, and Scombridae in the Gulf of Mexico, NOAA Technical Memorandum NMFS-SEFSC-317, 107 pp.

Klawe, W.L. and B.M. Shimada. 1959. Young scombroid fishes from the Gulf of Mexico. Bulletin of Marine Science of the Gulf and Caribbean 9: 100-115.

Lang, K.L., C.B. Grimes and R.F. Shaw. 1994. Variations in the age and growth of yellowfin tuna, *Thunnus albacares*, collected about the Mississippi River plume. Environmental Biology of Fishes 39: 259-270.

Leak, J.C. 1981. Distribution and abundance of carangid fish larvae in the eastern Gulf of Mexico, 1971-1974. Biological Oceanography 1: 1-28.

Lee, D.W., C.J. Brown, A.J. Catalano, J.R. Grubich, T.W. Greig, R.J. Miller, and M.T. Judge. 1994. SEFSC pelagic longline observer program data summary for 1992-1993. NOAA Technical Memorandum NMFS-SEFSC-347, 17p.

LGL Ecological Research Associates, Inc. and Science Applications International Corporation (SAIC). 1993. Cumulative ecological significance of oil and gas structures in the Gulf of Mexico: Information search, synthesis, and ecological modeling; Phase I, Final Report. U.S. Dept. of the Interior, U.S. Geological Survey, Biological Resources Division, USGS/BRD/CR--1997-0006 and Minerals Management Service, Gulf of Mexico OCS Region, New Orleans LA, OCS Study MMS 97-0036. vii + 130 pp.

Manooch, S.C., III., D.L. Mason and R. Nelson. 1984. Food and gastrointestinal parasites of dolphin, *Coryphaena hippurus*, collected among the southeastern and Gulf coasts of the United States. Bulletin of the Japanese Society of Scientific Fisheries 50: 1511-1525.

Manooch, S.C., III. and W.T. Hogarth. 1983. Stomach contents and giant trematodes from wahoo, *Acanthocybium solanderi*, collected along the South Atlantic and Gulf Coast of the United States. Bulletin of Marine Science 33: 227-238.

Manooch, S.C., III and D.L Mason. 1983. Comparative food studies of yellowfin tuna, *Thunnus albacares*, and blackfin tuna, *T. atlanticus*, from the southeastern and Gulf Coast of the United States. ACTA Ichthyologica et Piscatoria 8: 25-46.

Mather, F.J. III. 1962. Tunas (Genus *Thunnus*) of the Western North Atlantic: Part III Distribution and behavior of *Thunnus* species. Symposium on scombroid fishes, Mandapam Camp, S. India, Marine Biological Association of India, pp. 411-426.

Mather, F.J. III, J.W. Mason and A.C. Jones. 1995. Historical Document: Life history and fisheries of Atlantic bluefin tuna. US Dept of Commerce, NOAA, NOAA Technical Memorandum NMFS-SEFSC-370, 146 p.

Matthews, F.D., D.M. Damkaer, L.W. Knapp, and B.B. Collette. 1977. Food of Western North Atlantic tunas (*Thunnus*) and lancetfishes (*Alepisaurus*). NOAA Technical Report No. NMFS SSRT-706.

Maul, G.A, F. Williams, M. Roffer, and F. Souza. 1984. Remotely sensed oceanographic patterns and variability in bluefin tuna catch in the Gulf of Mexico. Oceanology Acta, 7: 469-479.

McGowan, M.F. and W.J. Richards. 1986. Distribution and abundance of bluefin tuna (*Thunnus thynnus*) larvae in the Gulf of Mexico in 1982 and 1983 with estimates of the biomass and population size of the spawning stock for 1977, 1978 and 1981-1983. Collective Volume of Scientific Papers- International Commission for the Conservation of Atlantic Tuna 24: 182-195.

McGowan, M.F. and W.J. Richards. 1989. Bluefin tuna, *Thunnus thynnus*, larvae in the Gulf Stream off the southeastern United States: satellite and shipboard observations of their environment. Fishery Bulletin 87: 615-631.

McKenney, T.W., E.C. Alexander and G.L. Voss. 1958. Early development and larval distribution of the carangid fish, *Caranx crysos* (Mitchill). Bulletin of Marine Science 8: 167-200.

National Ocean Service (NOS). 1985. Gulf of Mexico Coastal and Ocean Zones Strategic Assessment: Data Atlas. Strategic Assessment Branch, Ocean Assessments Division, Office of Oceanography and Marine Assessment, National Ocean Service, and the Southeast Fisheries Center, National Marine Fisheries Service, National Oceanic and Atmospheric Administration.

Palko, B. J., G. L. Beardsley and W. J. Richards. 1982. Synopsis of the biological data on dolphin-fishes *Coryphaena hippurus* Linnaeus and *Coryphaena equiselis* Linnaeus. FAO Fisheries Synopsis No. 130, 28 p.

Parr, A. 1939. Quantitative observations on the pelagic *Sargassum* vegetation of the western North Atlantic. Bulletin Bingham Oceanographic Collection 6: 1-94.

Pattillo, M.E., T.E. Czapla, D.M. Nelson and M.E. Monaco. 1997. Distribution and abundance of fishes and invertebrates in Gulf of Mexico estuaries. Volume II: Species life history summaries. ELMR Rep. No. 11. NOAA/NOS Strategic Environmental Assessments Divison, Silver Spring, MD. 377 p.

Power, J.H. and L.N.J. May. 1991. Satellite observed sea-surface temperatures and yellowfin tuna catch and effort in the Gulf of Mexico. Fisheries Bulletin 89: 429-439.

Price, J.M. and C.F. Marshall. 1996. Oil-spill risk analysis: central western Gulf of Mexico outer continental shelf lease sales 166 & 168. U.S. Dept. of the Interior, Minerals Management Service, Herndon, VA. OCS Report MMS 96-0013.

Richards, W.J. 1975. Spawning of bluefin tuna (*Thunnus thynnus*) in the Atlantic ocean and adjacent areas. Collective Volume of Scientific Papers- International Commission for the Conservation of Atlantic Tuna 5: 267-278.

Richards, W. J., M. F. McGowan, T. Leming, J. T. Lamkin and S. Kelley. 1993. Larval Fish Assemblages at the loop current boundary in the Gulf of Mexico. Bulletin of Marine Science 53: 475-537.

Richards, W. J., T. Leming, M. F. McGowan, J. T. Lamkin and S. Kelley-Fraga. 1989. Distribution of fish larvae in relation to hydrographic features of the loop current boundary in the Gulf of Mexico. Rapports et Proces-verbaux des Réunions. Conseil International Pour L'exploration De La Mer 191: 169-176.

Safina, C. 1993. Bluefin tuna in the west Atlantic: negligent management and the making of an endangered species. Conservation Biology, 7: 229-233.

Scott, G.P., S.C. Turner, C.B. Grimes, W.J. Richards and E.B. Brothers. 1993. Indices of larval bluefin tuna, *Thunnus thynnus*, abundance in the Gulf of Mexico; Modeling variability in growth, mortality, and gear selectivity. Bulletin of Marine Science 53: 912-929.

Settle, L. 1993. Spatial and temporal variability in the distribution and adundance of larval and juvenile fishes associated with pelagic *Sargassum*. M.Sc. Thesis, Univ. North Carolina at Wilmington.

Shaw, R.F. and D.L. Drullinger. 1990. Early life history profiles, seasonal abundance, and distribution of four species of Carangid larvae off Louisiana, 1982 and 1983. NOAA Technical Report No. 89. NTIS number: PB90-266404.

Sherman, K., R. Lasker, W.J. Richards and A.W. Kendall, Jr. 1983. Ichthyoplankton and fish recruitment studies in large marine ecosystems. Marine Fisheries Review, 45: 1-25.

South Atlantic Fishery Management Council. 1998. Fishery management plan for pelagic *Sargassum* habitat of the south Atlantic region. South Atlantic Fishery Management Council, 89 p.

Southeast Fisheries Science Center. 1992. Status of Fishery Resources off the Southeastern United States for 1991. NOAA, NOAA Technical Memorandum NMFS-SEFSC-306, 75 p.

Staiger, J. 1965. Atlantic flyingfishes of the genus *Cypseurus*, with descriptions of the juveniles. Bulletin of Marine Science 15: 672-725.

Uchida, K.S., S. Imai, S. Mito, M. Fujita, Y. Ueno, T. Shojima, T. Senta, M. Tahuka and Y. Dotsu. 1958. Studies on the eggs, larvae, and juveniles of Japanese fishes. Series 1: Second laboratory of Fisheries Biology. Fisheries Department, Faculty of Agriculture, Kyushu University, Fukuoka, Japan.